Smart Materials

Also of Interest

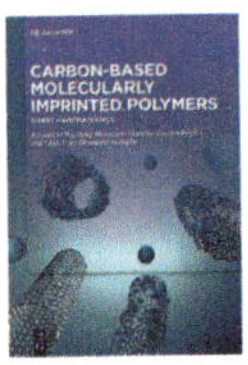

Carbon-Based Molecularly Imprinted Polymers.
Smart Nanomaterials
Chaudhery Mustansar Hussain, Rüstem Keçili,
Chaudhery Ghazanfar Hussain (Eds.), 2025
ISBN 978-3-11-150302-8
e-ISBN (PDF) 978-3-11-150332-5

Advanced Biomedical Composites.
Materials, Design, Manufacturing
De Gryuter Series of Advanced Composites
Sumit Pramanik, Sandipan Roy, J. Paulo Davim (Eds.), 2025
ISBN 978-3-11-138594-5
e-ISBN (PDF) 978-3-11-138602-7

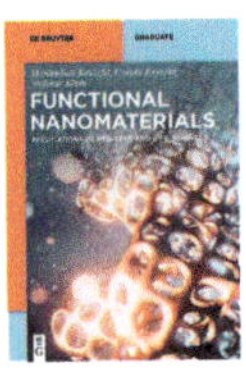

Functional Nanomaterials.
Applications in Medicine and Life Sciences
Stefanie Klein, Maximilian Kryschi, Carola Kryschi, 2024
ISBN 978-3-11-060545-7
e-ISBN (PDF) 978-3-11-060549-5

Smart and Functional Textiles
Bapan Adak, Samrat Mukhopadhyay (Eds.), 2023
ISBN 978-3-11-075972-3
e-ISBN (PDF) 978-3-11-075974-7

Smart Materials

Biomaterials, Coatings, Energy, Sensors and Structural Engineering

Edited by
Ajay Kumar, Parveen Kumar, Tamer Sebaey
and Velaphi Msomi

DE GRUYTER

Editors
Prof. Dr. Ajay Kumar
Department of Mechanical Engineering,
School of Core Engineering, Faculty of Science,
Technology & Architecture, Manipal University,
Jaipur, India.
ajay.kumar30886@gmail.com

Parveen Kumar
Skill Department of Automotive Studies,
Shri Vishwakarma Skill University, Palwal,
Haryana, India
parveenkumarme@gmail.com

Prof. Dr. Tamer Sebaey
Department of Production and Manufacture
Engineering
College of Engineering
Prince Sultan University
Riyadh 11586, Saudi-Arabia
tsebaey@psu.edu.sa

Prof. Dr. Velaphi Msomi
Associate Professor
Department of Mechanical Engineering
University of South Africa
Preller Street
Pretoria 0002, South Africa
msomiv@gmail.com

ISBN 978-3-11-151283-9
e-ISBN (PDF) 978-3-11-151571-7
e-ISBN (EPUB) 978-3-11-151689-9

Library of Congress Control Number: 2025943710

Bibliographic information published by the Deutsche Nationalbibliothek
The Deutsche Nationalbibliothek lists this publication in the Deutsche Nationalbibliografie; detailed bibliographic data are available on the internet at http://dnb.dnb.de.

Cover image: Alllex/iStock/Getty Images Plus
Typesetting: Integra Software Services Pvt. Ltd.
Printing and binding: CPI books GmbH, Leck

www.degruyterbrill.com
Questions about General Product Safety Regulation:
productsafety@degruyterbrill.com

Preface

Smart materials have garnered significant attention in recent years due to their unique properties and promising applications across various industries. Smart materials find extensive use in diverse fields such as aerospace, healthcare, energy, and electronics. They enable the development of adaptive structures, self-healing materials, sensors, actuators, and energy harvesting devices, revolutionizing various industries and driving innovation. These advancements pave the way for smarter, more efficient technologies, shaping the future of materials science.

Smart materials are the materials which encompass properties that can be altered in a controlled manner by external or internal stimuli such as temperature, stress, pressure, pH, electric or magnetic fields, and moisture. The smart materials are also termed as intelligent materials because of their intellectual behavior in response to environmental conditions. The key features of intellectual behavior include the capabilities to retain original state once stimulus has been removed. The properties of smart materials such as self-sensing, self-healing, memory capabilities, self-adaptivity, self-diagnostics, self-actuating enable manifold applications in mechanical, electronics, medical, healthcare, aerospace, aviation, automobile, and environmental engineering. Smart materials have already found applications in various sectors but the development and innovation are still in its infancy. So the focus of this book is to gather information across the globe from all stakeholders of smart materials.

Advances in smart materials from different engineering applications and how these materials interact with the environment have a perspective of this book. This book will capture the recent innovative breakthroughs in smart materials to understand the broad series of themes ranging from experimental and computational analysis to hybrid advancement in high-tech intelligent materials, fundamental practices, and the latest developments in the applications of smart materials.

The book consists of 13 chapters that describe the perspectives of smart materials:

- Chapter “Smart Surfaces and Coatings: An Innovative Approach” describes various types of smart surfaces and coatings, including advanced superhydrophobic, anticorrosion, and antibacterial ones.
- Chapter “Emerging Technologies in Cementitious Material Design: Autonomous Healing and Durability Enhancement Mechanisms” discusses the most recent developments of self-healing systems with particular emphasis on their working principles, efficiency, and real-world usability.
- Chapter “A Study on the Strength and Stiffness of Shape Memory Alloy-Reinforced Structural Composites” investigates using shape memory alloys (SMAs) in laminated composites to improve structural stability, stiffness, and frequency responses under thermal loading via ABAQUS simulation and validation tests.
- Chapter “Modelling Techniques to Analyse the Functionality of Metallic Glass Alloys as Biomedical Implants: A Mini-review” outlines the importance of imple-

https://doi.org/10.1515/9783111515717-202

menting various modeling techniques such as finite element analysis (FEA), molecular dynamic (MD) simulations, statistical modeling, and machine learning to study mechanical properties and glass-forming ability of various compositions, which would make the development of novel metallic glass alloys economical and efficient for biomedical applications.
- Chapter “Numerical Modelling of URM Piers of Various Bond Patterns” presents results on the response of a two-dimensional masonry wall of various bond types with different aspect ratios subjected to in-plane lateral pushover loadings. The model is simulated by varying the interface properties of brick such as crushing, elastic, cracking, and shear properties.
- Chapter “Sustainability and Durability Synergy: Self-Healing Concrete in Ultra-High-Rise Structures” combines experimental approaches and numerical modeling to investigate the behavior of self-healing ultra-high-performance concrete in corrosive conditions.
- Chapter “Experimental Investigations on Metallic Current Collectors on the Performance of PDMFC” discusses current collector materials, such as brass, stainless steel, and nickel, having methanol compatibility and high electrical conductivity, which are chosen for experimental investigation using a single PDMFC.
- Chapter “In-Plane Dynamic Crushing Behavior of Hierarchical Honeycomb Sandwich Panel Under Blast Loading: A Finite Element Study” investigates the in-plane dynamic crushing behavior of hierarchical honeycomb sandwich panels under blast loading using Abaqus/Explicit. The panel demonstrated the superior energy absorption, reduced back-sheet deflection, and enhanced shock mitigation.
- Chapter “Utilizing the U-Net Image Machine Learning Segmentation Model for Automated Analysis of Grain Sizes in Metallographic Microstructures” presents a novel application of the U-Net deep learning model for accurate grain size measurement and segmentation in complex metallographic images, demonstrating high performance with minimal training data and significant alignment with manual measurements.
- Chapter “Advancements in Banana Fiber-Reinforced Composites: Mechanical Properties, Sustainability, and Scientometric Analysis” presents a comprehensive analysis of banana fiber-reinforced composites, focusing on their mechanical and thermal properties, sustainability benefits, and scientometric trends through detailed literature and data evaluation.
- Chapter “Exploring the Applications of Smart Materials in the Medical and Healthcare Sectors” examines the capabilities of various smart materials in the healthcare and medical fields, as they possess biocompatibility and biodegradability, leveraging these characteristics, and these materials are highly suitable for biomedical applications.
- Chapter “Advancements in 3D Printing of Composite Materials: Challenges and Prospects” presents an in-depth study of 3D printing of composite materials, focusing on technological fundamentals, continuous fiber reinforcement, hybrid

manufacturing, and emerging sustainable solutions for structural and functional advancements.
- Chapter "Colloidal Formulation and Robocasting of Silicon Nitride Scaffolds with Tailored TPMS Structures" discusses about a novel Si_3N_4 colloidal ink, incorporating polyvinyl alcohol, carboxymethyl cellulose, polyethylene glycol, and hydroxypropyl methylcellulose, which was developed with a high solid loading of approximately ~42 vol% and optimized for DIW by tuning its electrokinetic and rheological properties, followed by refinement of process parameters to achieve an effective scaffold structure.

This book will help readers pursue contemporary insights into currently ongoing practices in the fabrication of shape memory alloys, shape memory polymers, piezoelectric materials, electrochromic materials, thermoelectric materials, and magnetorestrictive materials. The book will empower and solve problems for academic scholars, manufacturing and material scientists, and commercial manufacturers of smart materials.

The editors acknowledge the professional support received from De Gruyter and express their gratitude for this opportunity.

Reader's observations, suggestions, and queries are welcome.

Editors
Ajay Kumar
Parveen Kumar
Tamer Sebaey
Velaphi Msomi

Acknowledgments

The editors are grateful to De Gruyter for showing their interest to publish this book in the buzz area of smart materials. The editors express their personal adulation and gratitude to Hélène Chavaroche, Acquisition Editor in Materials Science, De Gruyter, for giving consent to publish our work. She undoubtedly imparted the great and adept experience in terms of systematic and methodical staff who have helped the editors to compile and finalize the manuscript. The editors also extend their gratitude to Marie Hammerschmidt (Content Editor), De Gruyter, for supporting during this tenure.

Editors are thankful to the researchers, experts, and interviewees who generously shared their knowledge and experiences. Their contributions have enriched the content and added depth to the ideas presented in this book.

Dr. Ajay also wishes to express his gratitude to his parents, Sh. Jagdish and Smt. Kamla, and his loving brother Sh. Parveen for their true and endless support. They have made him able to walk tall before the world regardless of sacrificing their happiness and living in a small village. He cannot close these prefatory remarks without expressing his deep sense of gratitude and reverence to his life partner Mrs. Sarita Rathee for her understanding, care, support, and encouragement to keep his moral high all the time. No magnitude of words can ever quantify the love and gratitude he feels in thanking his daughters, Sejal Rathee and Mahi Rathee, and son Kushal Rathee, who are the world's best children.

Finally, the editors obligate this work to the divine creator and express their indebtedness to the "ALMIGHTY" for gifting them power to yield their ideas and concepts into substantial manifestation. The editors believe that this book would enlighten the readers about each feature and characteristics of smart materials.

Ajay Kumar
Parveen Kumar
Tamer Sebaey
Velaphi Msomi

https://doi.org/10.1515/9783111515717-203

Contents

Amandeep Singh*, Sarthik Samanta, and Santosh K. Tiwari

Smart Surfaces and Coatings: An Innovative Approach

Abstract: Smart surfaces and coatings have emerged as a significant trend in surface technology in recent time. Their possible applications are varied and hold the promise of transforming numerous industries. These are advanced or intelligent materials that respond dynamically to external stimuli, thereby offering benefits that extend beyond the fundamental, passive roles of traditional coatings, such as decoration and protection. These stimuli may include heat, light, mechanical forces, temperature changes, pressure variations, pH shifts, and exposure to corrosive ions. As coating technology continues to evolve, these materials have become essential in industrial manufacturing, exhibiting remarkable properties such as superhydrophobicity, self-cleaning capabilities, enhanced antibacterial characteristics, and improved resistance to corrosion. Biomimetics, the study of the structures, functions, and behaviors of living organisms, has a decisive part in engineering design. The integration of biomimetic principles with intelligent coating materials can enhance their performance and functionality, foremost to the advancement of even more advanced coatings. In this chapter, various types of intelligent coatings including advanced superhydrophobic coatings, anticorrosion coatings, and antibacterial coatings are discussed in detail.

Keywords: Advanced materials, intelligent materials, biomimetics, superhydrophobicity, self-healing

1 Introduction

Smart surfaces and coatings signify a transformative advancement in coating expertise, integrating nanomaterials along with intelligent functionalities in order to produce coating system with remarkable characteristics. This chapter examines the significant progress and potential uses of smart surfaces and coatings across diverse

*Corresponding author: Amandeep Singh, Research Scientist, India Institute of Technology (IIT), Ropar 140001, Punjab, India, e-mail: adsingh.chemical@gmail.com
Sarthik Samanta, Earth Science and Engineering, Missouri University of Science and Technology, Missouri, Rolla, MO 65409, USA
Santosh K. Tiwari, State Key Laboratory of Featured Metal Materials and Life-Cycle Safety for Composite Structures and MOE Key Laboratory of New Processing Technology for Nonferrous Metals and Materials, School of Resources, Environment and Materials, Guangxi University, Nanning 530004, China; Centre for New Materials and Surface Engineering, NMAM Institute of Technology, Karnataka, Nitte 574110, India

https://doi.org/10.1515/9783111515717-001

industries. It investigates the evolution of antifouling as well as self-cleaning coatings, emphasizing the role of nanotechnology in fostering innovations within these domains. Additionally, this chapter underscores the importance of intelligent coatings which adjust to the environmental changes and other various external stimuli, illustrating the possible applications in energy-efficient construction constituents, aerospace technologies, and medical-related tools. It reviews new developments in the stimuli-responsive smart coating materials and their capacity to revolutionize the industries through enhancing their working and promoting sustainability aspects as well. The chapter will then be briefed by discussing various challenges in this domain along with prospective opportunities associated with smart coatings including issues related to cost-effectiveness, scalability, as well as ecological impact. It highlights the crucial part of smart surfaces and coatings expected to important in the advancement of materials science and technology, flagging the further way for effective as well as sustainable resolutions.

2 Methods for Coating Process

2.1 Galvanization for Metal Surfaces

The galvanization represents the predominant method of surface coating widely utilized across various industrial sectors. The fundamental principle of galvanization is a galvanic shield to underlying the steel substrate. In the natural world, various elements position based on their tendencies toward reduction or oxidation. An essential framework known as the EMF series is established, which quantifies the potential differences between hydrogen and various metals on an individual basis. Consequently, noble metals position at upper end of EMF series, while reactive metals place at lower end. The highly reactive metals exhibit a greater propensity for corrosion compared to their noble counterparts, attributable to their significant potential differences. If two dissimilar metals interconnect, the rate of corrosion of highly unreactive metal diminishes, whereas that of the more reactive metal escalates. In this manner, the reactive metal serves to protect unreactive metal from the corrosion process by reducing the rate of corrosion. Within the context of the galvanizing, base steel functions as the noble metal, while the zinc coating serves as the active metal. Should the zinc layer sustain mechanical damage or scratches, it behaves as anode, with base steel acting as cathode, thereby reducing corrosion rate of base steel and providing it with protective benefits against corrosion. Various methods are mentioned in Fig. 1.

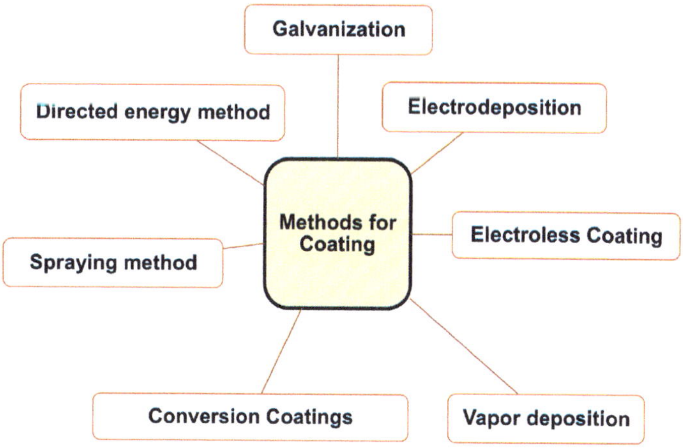

Fig. 1: Conventional method for the coating process.

2.2 Electrodeposition Method

The method of electrodeposition for surface coating is widely utilized not only for the application of coatings on the surfaces of metal but also for metal extraction. This process, commonly referred to as electroplating, is fundamentally based on Faraday's law. According to Faraday, mass of a substance that is deposited/released from any surface remains directly proportional to atomic mass as well as quantity of the current that flows through. Electroplating setups typically include cathodes, anodes, along with electrolytes. The direct current (DC) is employed to facilitate the release/deposition of ions. Deposition is achieved through various processes, including electrorefining, electrowinning, molten salt techniques. The various metal ions are deposited onto base metal that serves as cathode for a particular electrochemical cell. During refining procedure, the impure metal acts as the anode while base metal functions as the cathode. Since impure metal is polarized anodically through the internal DC, it leads to the release of the desired metal ions.

2.3 Electroless Coating Method

Main distinction between electroplating and the electroless coating lies in lack of outer power in case of electroless coating. In electroless coating, the initial step involves cleaning the target surface, which is then submerged in a chemical solution. Furthermore, ions present in solution interact with those on surface, leading to etching. Additionally, the ions intended for the coating are substituted for the etchant ions due to the potential difference. This technique is commonly employed to deposit nickel atoms onto surfaces.

2.4 Vapor Deposition (VD) Techniques

Vapor deposition (VD) techniques employ the processes of vaporization and condensation to create thin layers. Initially, selected metal is subjected to evaporation followed by subsequent condensation onto substrate. These VD techniques are categorized into two primary types: physical vapor deposition (PVD) and chemical vapor deposition (CVD).

2.4.1 PVD Method

This technique has potential to develop an ultra-thin film having thickness up to 20 µm. Processes of atomization/ionization is achieved through simply evaporating the material or using plasma sputtering method. In case of simple evaporation approach, metallic material is heated to the elevated temperatures, typically up to 2,000 °C, followed by subsequent condensation in the vacuum condition. Conversely, the plasma sputtering technique involves the emission of coating material from the surface by directing accelerated ions to the target substance. The accelerated ions strike to surface, resulting the ejection of anticipated atoms of coating material. These atoms are then deposited onto intended substrate. The operation occurs within an electrostatic framework, where the target material functions as cathode. Since the argon gas serves as medium, wherein this gas is subjected to be ionized, and the resulting ions propelled by metallic platform, facilitating the ejection of atoms from its surface. Application of plasma sputtering technique significantly enhances the adhesion capacity between substrate and coating material due to the involvement of high kinetic energy.

2.4.2 CVD Method

This technique involves utilizing gaseous form of target metal that is subsequently condensed on surface of the base metal. Substrate is subjected to elevated heating exceeding 850 °C. Such elevated temperature conditions limit range of materials that can be employed, as only those with elevated melting points are suitable for this method. Initially, precursor materials are subjected to heated chamber containing target surface. These precursors undergo vaporization and then adsorbed on target surface. The adsorbed materials engage in chemical reactions, resulting in deposition of desired metal coating on surface. This process yields a dense, thin layer of target metal on base material. This coating technique is frequently applied in deposition of various hard materials.

2.5 Conversion Coatings for Metal Surfaces

The conversion coatings are typically applied through spraying or dipping the sub strate in a specific solution. These coatings enhance corrosion resistance of surface by utilizing extra substances. Furthermore, conversion coatings improve surface's adhesion capabilities, facilitating subsequent painting processes. The methods for applying conversion coatings are categorized into phosphating, chromating, and anodizing. Phosphating process consists of the immersion of metal surfaces in phosphoric acid solution, resulting in the formation of a phosphate layer that enhances adhesion for subsequent paint applications. The chromating includes the immersion of substrate in solution-containing chromium ions. The chromate coatings exhibit self-healing characteristics because chromium is gradually released upon exposure to moisture or liquid. The leached hexavalent chromium ions are absorbed by surface defects, ultimately leading to the formation of an inactive film composed of trivalent chromium which are insoluble in nature. The anodizing process differs from chromating and phosphating in that protective film is generated by base metal itself. Certain metals, such as titanium and aluminum, can develop a shielding layer of oxide on surfaces.

2.6 Spraying Method

2.6.1 Thermal Spraying Method

In this method of surface coating, materials are subjected to heating and subsequently sprayed onto the targeted metal surface. This technique allows for a coating thickness that can exceed that achieved by other conventional methods. Typically, the heating of these materials is facilitated by an external electrical source capable of producing either an arc or plasma. Thermal spraying encompasses a range of coating processes wherein materials that are melted are applied to the surface. The precursor for the coating is further heated through either electrical means (using plasma method or arc) or through any chemical processes. This technique is notable for its ability to produce thick coatings, with thicknesses ranging from approximately up to several millimeters, depending on specific procedure and precursor used. Furthermore, thermal spraying allows for the application of coatings over extensive areas at a higher deposition rate compared to other methods including electroplating and both PVD and CVD. The materials suitable for thermal spraying encompass a variety of options including various metals, their alloys, different ceramics, various plastics, and their composites. These materials are supplied in the form of powder or wire form. Flame powder and plasma coating method are shown in Fig. 2.

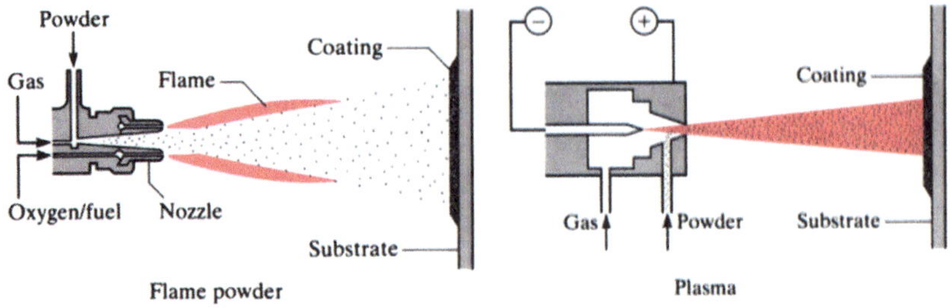

Fig. 2: Flame powder and plasma coating method. Reproduced from [1], Creative Commons BY-NC-SA 4.0: The Open University.

2.6.2 Cold Spraying

Gas dynamic cold spraying, commonly referred to as cold spraying, is actually a method used for coating deposition. This process includes the propelling of solid powders with diameters ranging from 1 to 50 µm within the gas jet, achieving speeds of approximately 1,200 m/s. Upon colliding with substrate, these units experience plastic deformation, resulting in their adhesion to surface. In order to ensure a consistent coating width, spraying nozzle is maneuvered across the surface of substrate. This technique allows for the deposition of various materials. Kinetic energy imparted to particles through gas expansion is transformed into plastic deformation energy during the bonding process. In contrast to thermal-spraying methods, powders remain in a solid state throughout the spraying operation.

2.6.3 Plasma Method

The plasma spraying method falls under the thermal-coating approach which uses elevated heat and high-energy electrically generated heat sources to melt and propel fine particles onto a substrate. This process encompasses the creation of a plasma jet, the introduction and processing of particles within this jet, and ultimately the development of the coating. Upon contact with the substrate, the molten metal particles solidify, resulting in the formation of uniform coatings.

2.7 Surface Modification by Directed Energy Techniques

A substantial development in domain of materials technology is the adoption of advanced techniques for managing composition and microstructure throughout the processing stages. Among these innovative approaches, near-surface modifications through ion implantation and laser treatment stand out as effective methods for customizing material characteristics. These techniques have led to the creation of novel materials that exist far from thermodynamic equilibrium, displaying unexpected yet advantageous properties. The alterations in properties that have garnered the most attention include enhanced electrical characteristics in semiconductors as well as increased wear resistance, hardness, and corrosion resistance in metals. The precise control afforded by energetic beams facilitates a systematic exploration of the relationships between microstructure and material properties at the atomic scale.

2.8 Thermochemical Treatment

The thermochemical treatment method uses thermal diffusion property in order to introduce nonmetallic/metallic atoms to surface of the material, thereby altering the chemical properties and microstructure. This process can occur in all the phases, often involving various active species simultaneously. Typically, mechanism of thermochemical treatments encompasses decomposition of the elements, the dissociation of molecules to generate the nascent atoms, the simultaneous absorption of produced atoms, followed by subsequent diffusion, and the reactions that occur in substrate to either alter the phases or create another one. In engineering applications, where entire component exposes to elevated heating, the surface diffusion interacts with various changes occurring in volume of material, which may include phase transformations, thereby increasing the complexity of the process. Additionally, thermochemical modifications encompass various techniques which also investigate elements like V and Mo.

2.9 Wet chemical

Wet coating represents a widely adopted technique for applying coatings across a diverse array of substrates. This spraying method facilitates the achievement of various effects, surface textures, and color variations. The pretreatment of wet chemical method remains integral to pretreatment processes of surface in numerous manufacturing divisions including therapeutic science, aerospace, and locomotive manufacturing. The Fraunhofer Institute for Manufacturing Engineering and Automation contributes to this domain by developing and refining a variety of procedures and required products aimed at this method. The application of wet chemistry offers significant rewards for the pretreatment of metals. This approach serves multiple functions, including cleaning (elimination of contaminants from prior processing), etching, and pickling (elimination of undefined layers, adjustment of roughness of the surface, and gloss levels) as well as protection and activation through various methods. Furthermore, wet chemistry provides distinct benefits over mechanical or physical pretreatment methods for metals such as Al, Mg, Ni, and Ti, which are critical for certain applications. Additionally, wet chemistry can treat or modifying areas that are inaccessible to techniques like laser or plasma technology such as internal cavities or intricate structures. This capability is particularly relevant in the context of metal 3D printing [2].

2.10 Electrochemical processes

The electrochemical processes for the surface finishing represent a scalable industrial technique which conventionally employs acidic electrolytes in order to attain specific surface properties on the metallic components. This traditional method often incorporates aggressive and dangerous chemical agents to eliminate oxide layer from materials that exhibit strong passivity. A novel strategy has emerged that utilizes pulsed and pulsed opposite electric fields to regulate the distribution of current, reduce the creation of oxide films, and attain the anticipated surface, all while utilizing environmentally friendly and straightforward chemical compositions. This innovative approach not only reduces manufacturing costs but also enhances the overall robustness of the process.

3 Properties and Performance of Smart Surfaces and Coatings

3.1 Antifriction Performance

Antifriction coatings function as dry lubricants that effectively minimize friction and wear. Their ability to mitigate noise makes them versatile for application across various materials, indicating that these coatings are not limited to steel and aluminum. Coatings made from polymers and fluoropolymers have demonstrated significant efficacy in ensuring that steel components possess excellent nonstick and sliding characteristics, along with enhanced wear resistance. Utilizing a specialized infusion technique, the high-performance plastic surface is securely bonded to the substrate. To further enhance wear protection, we incorporate additional interlayers that provide enduring defense against wear and oxidation. This treatment is not only highly advisable for steel and stainless steel but also offers optimal sliding and nonstick properties for materials such as aluminum, copper, ceramics, and plastics. Main performance of smart surfaces and coating materials is mentioned in Fig. 3.

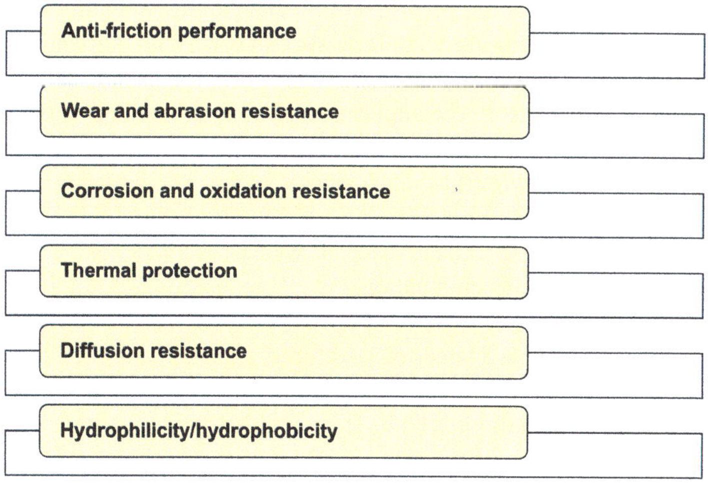

Fig. 3: Main performance of smart surfaces and coatings.

3.2 Wear and Abrasion Resistance

Wear-resistant coatings serve the primary function of safeguarding materials from wear and tear. These protective layers contribute to the preservation of performance standards, thereby prolonging the lifespan of structural components. The wear experienced by a structure is often a result of heightened friction between interacting surfaces, and these coatings are specifically designed to shield those surfaces from potential damage. Research indicates that mechanical aspects including hardness of such coating materials are significantly enhanced through the integration of nanoparticles, including silicon SiO_2, TiO_2, Al_2O_3, and ZrO_2 [3]. Enhanced mechanical properties lead to increased resistance to wear and scratching. Additionally, the diminutive size of nanoparticles contributes to their ability to preserve the transparency and gloss of building coatings. Consequently, these nano-modified coatings can serve dual functions: enhancing durability while simultaneously maintaining the aesthetic qualities of treated surfaces such as floors and window panes. Carbide coatings are utilized to protect surfaces from erosion, abrasion, fretting, and galling. These coatings are often applied using advanced thermal spray techniques, although plasma coating is also an option. This application process significantly improves the material properties of the coated surfaces, enhancing hardness, maintaining material integrity, increasing resistance and durability, and safeguarding metals from adhesive wear. A variety of carbide coatings are available including tungsten carbide-cobalt/chrome, tungsten carbide-cobalt, tungsten carbide/chromium carbide-nickel, tungsten carbide-nickel, and chromium carbide-nickel/chrome, among others [4]. The ceramic coating materials, often referred to as oxide coatings, serve to safeguard components from various forms of wear including abrasion, sliding wear, fretting, and galling. In addition to their protective qualities, these coatings provide resistance to corrosion, particularly when used in conjunction with sealers, making them suitable for applications where exposure to corrosive fluids may occur. The application of ceramic coatings is typically achieved through techniques such as plasma spraying or detonation gun methods. A diverse range of ceramic coatings is available, including but not limited to aluminum oxide, titanium oxide, chromium oxide, and combinations thereof, such as chromium oxide with titanium oxide or silicon oxide, as well as zirconium oxide and yttrium oxide, either individually or in combination.

Cermet coatings are engineered to function similarly to ceramic and carbide coatings, but they operate effectively at elevated temperatures, approximately 1,100 °C. The binder utilized in these coatings is enhanced with superalloys MCrAlY, where "M" can represent Co, Ni, or the combination of both. This modified binder exhibits improved oxidation resistance at higher temperatures. Various compositions include MCrAlY/Cr_3C_2, superalloy/Al_2O_3, MCrAlY/Al_2O_3, NiCrBSi/WC carbide, and blends of various superalloy with oxides or carbides. The metal alloy coating is based on a foundational alloy that demonstrates corrosion resistance while enduring multiple challenging wear conditions. A range of thermal spraying techniques can be employed to apply these metal alloy coatings including combinations such as Ni superalloy/WC, Ni superalloy/Cr_3C_2, and Co alloys/WC. The laser cladding offers a metallurgical bond with predictable chemistry and ductility, effectively addressing wear and corrosion in applications where thermal spray coatings or chrome plating may fall short due to issues like brittleness, infusible morphology, or weak adhesive bonds. Examples of such applications include Ni alloys/WC, stainless steel/WC, Ni-based hard-facing alloys, WC composites, and Co-based hard-facing alloys.

3.3 Corrosion and Oxidation Resistance

Corrosion-resistant coatings serve a crucial role in mitigating the deterioration caused by factors such as oxidation, moisture, and chemical exposure. These coatings enhance the protection of metallic components by acting as a barrier that inhibits direct contact with corrosive agents and chemical substances. Specifically, corrosion-resistant coatings are designed to safeguard metal surfaces from corrosion and can be formulated from materials like zinc or iron. They may be applied in various forms including base layers, primers, or topcoats, utilizing a range of application techniques. Such coatings are frequently employed to protect infrastructure elements, including pipelines, and reinforcing bars in concrete. Additionally, some formulations incorporate anticorrosive pigments or corrosion inhibitors, which further enhance the protective capabilities for metal substrates including zinc, steel, and aluminum.

3.4 Thermal Protection

A thermal barrier coating (TBC) is a specialized, thin layer applied to various materials to safeguard them against overheating, cracking, or degradation when subjected to extreme temperatures. These coatings function as a thermal insulator, effectively diminishing the rate of heat transfer to the underlying material. TBCs are sophisticated materials utilized on surfaces that encounter high thermal conditions, specifically engineered to insulate, and shield the substrate from intense heat exposure. Typically composed of a ceramic material, such as zirconia, these coatings exhibit low thermal conductivity, thereby serving as an effective barrier that minimizes heat exchange with the substrate. TBCs find extensive application in aerospace sectors, particularly in turbine blades and engine components, as well as in industrial gas turbines and automotive engines, where they contribute to enhanced efficiency and prolonged component lifespan. Additionally, TBCs are frequently employed in the sealing of building envelopes, which are the outer structures of buildings that delineate the interior from the external environment. The application of TBCs to building envelopes enhances insulation and boosts energy efficiency by curtailing heat transfer. By effectively sealing these envelopes, TBCs play a crucial role in maintaining a stable indoor climate, lowering energy consumption, and promoting the overall sustainability of the building.

3.5 Diffusion Resistance

Diffusion coating refers to a technique wherein metal components, intended for exposure to extreme temperatures and aggressive corrosive environments, are enveloped in a noncorrosive material. This procedure typically occurs at elevated temperatures within a controlled chamber. Surface of metal is altered through diffusion of the metal or element at high temperatures, which imparts the necessary resistance when integrated with the base metal. These coatings are designated as “diffusion coatings.” This method can be utilized on various metals and alloys, including nickel, titanium, and molybdenum, although it is most applied to ferrous metals. Notable examples include zinc diffusion coatings and aluminum diffusion coatings [5].

3.6 Hydrophilicity/Hydrophobicity

In the realm of surface technology, the utilization of hydrophilic and hydrophobic coatings presents significant advantages across a multitude of sectors. This chapter explores the differences, applications, and implications of these coatings, serving as a comprehensive guide to how each type contributes to the enhancement of material performance. Hydrophilic coatings are designed to attract water, rendering surfaces more amenable to moisture. By lowering surface tension, these coatings facilitate the uniform spreading of water, thereby improving wettability. On the other hand, hydrophobic coatings function by repelling water, establishing a protective barrier that inhibits water penetration, which results in the formation of moisture that bead on surface. The emphasis on hydrophilic surfaces highlights the function in fostering absorption and distribution of water, in contrast to hydrophobic coatings which opposes the moisture infiltration. Chemical structure of hydrophilic surfaces typically incorporates various polar natured functional groups which draw water. Such surfaces are specifically formulated to establish hydrogen bonds with the water molecule using various materials such as silica, polyethylene glycol, and zwitterionic types polymers. In contrast, hydrophobic surfaces are defined by their nonpolar molecular composition, including substances like fluoropolymers and silicones, which effectively repel water [6].

Since nanotechnology plays a vital role in almost every field of science, nanocomposites of various polymers have huge impact on the advancement of the research. Recently, MXene-based nanocomposites and various quantum dotsbased nanocomposites emerge as functional materials for different applications. Various natural polymer, biopolymers, and nanomaterials, such as hydroxyapatite, and cellulose, have potential to be used to develop advanced materials. Therefore, the development of smart surfaces and smart coating is also influenced by the nanomaterials and nanotechnology at the greater extent.

4 Applications of Smart Surfaces and Coatings

4.1 Bioinspired Coating

Nature consistently demonstrates its efficacy by providing solutions to a wide range of challenges, whether they are straightforward or intricate. A notable instance of inspiration drawn from nature is the development of bioinspired coatings. These innovative coatings possess remarkable features such as responsiveness to environmental stimuli, antimicrobial effectiveness, antifouling properties, self-healing abilities, and exceptional wettability. Such attributes have garnered the attention of researchers across various disciplines, fostering interdisciplinary collaboration. This discourse encompasses the scientific principles and engineering methodologies pertinent to the creation of intelligent biomimetic surfaces and coatings. Structural colors, particularly those that are angle-independent and arise from wavelength-dependent light scattering, have attracted substantial consideration due to the extensive applications in areas such as displays, sensors, and detection technologies. However, these colors often undergo substantial changes or may even vanish during practical use, a phenomenon attributed to variations in refractive index caused by liquid wetting or infiltration. Drawing inspiration from the feathers of birds, Cheng et al. [7] introduced a straightforward deposition method for creating a coating that exhibits angle-independent structural color and superamphiphobic properties as shown in Fig. 4. This coating features channel-type structures approximately 200 nm in size, formed between hollow silica and air nanostructures, and displays a resilient sapphire blue hue that remains unaffected by significant liquid intrusion, mirroring the properties of the back feathers of the Eastern Bluebird. The biomimetic coating achieves high color saturation and superamphiphobicity through the optimization of coating parameters and the incorporation of dark materials. Its exceptional durability in challenging conditions ensures a prolonged service life across various extreme environments.

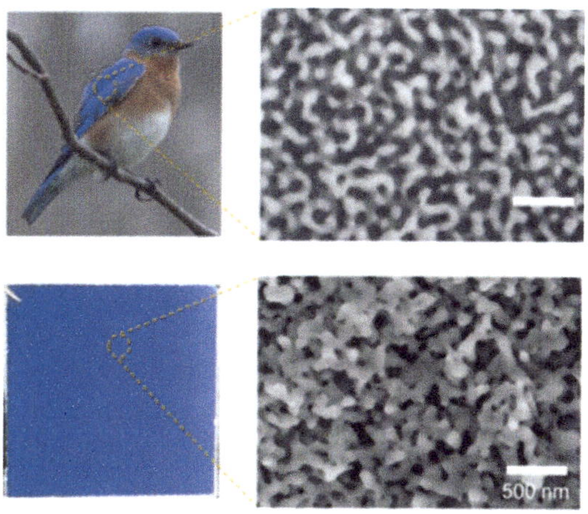

Fig. 4: Channel-type β-keratin and air nanostructure from the back contour feather barbs of eastern bluebird. Copyright © 2024 American Chemical Society, reproduced with due permission from [7].

In another work, anticorrosive properties of nacre-inspired GO/epoxy (NBGE) surfaces structured with the alternating layers were investigated [8]. Potentiodynamic tests demonstrated that rate of corrosion is decreased by coating of NBGE materials, which comprised GO and epoxy (5NBGE) multilayers of 17-μm thickness. The corrosion was reduced to one-twentieth of that observed in steel covered by a pristine coating of epoxy that was twice as thick, when immersed in a 3.5% salt solution. Performance of the 5NBGE materials surpassed that of carbon-based nanoparticle/epoxy composite-coating materials. Anticorrosion mechanism of NBGE coating materials is elucidated through chemically reduction at a steel-coating interface, attributed to limited penetration of oxygen water, and chloride ions facilitated by diminished defects due to intercalated layers of GO within coating surfaces. Furthermore, A novel bioinspired composite coating comprising octacalcium phosphate (OCP) and silk fibroin (SF) was successfully developed on Ti substrates using gentle deposition technique [9]. The results reveal that SF is indispensable for the formation of the distinctive biomimetic architecture of the composite of OCP/SF coating. The in vitro experiments indicate that coating of composite of OCP/SF, characterized by their organized and porous framework, significantly improves cellular responses, designated it appropriate for the range of hard tissue engineering uses. In the same way, a bioinspired material was developed by Li et al. [10], which exhibits a low coefficient of friction, even when subjected to extremely high loads, and demonstrates an exceptionally long lifespan. Additionally, the composite material exhibits zero-wear characteristics and inflicts minimal wear damage on the opposing friction surface. The enhanced tribological performance is ascribed to the formation of a spontaneously generated layer that mimics articular cartilage, which is continuously replenished through the application of frictional heat and pressure.

In another study, a slippery zinc phosphate coating characterized by uniform pore structures was successfully developed using a straightforward method inspired by the morphology of pitcher plants [11]. Coating showed superhydrophobic properties, achieving a significant water contact angle following treatment using a fluoroalkylsilane ethanol. The incorporation of Krytox 100 to pores resulted in a surface with exceptional water-repellent characteristics. Furthermore, this innovative slippery coating demonstrated remarkable anticorrosion capabilities, with anticorrosion properties improved by many folds than a simple steel substrate. Notably, coating maintained a corrosion inhibition efficiency exceeding 99.99% after 6 weeks of the immersion, indicating outstanding longstanding steadiness for the anticorrosion application. This coating holds considerable promise for applications in anti-icing uses, antibacterial uses, as well as for water harvesting technologies. In another work, a photoelectronic antibacterial as well as antifogging coating material was developed, drawing inspiration from catfish and sphagnum moss [12] as shown in Fig. 5. This coating was created through the cross-linking of polyethylenimine (PEI)-modified titanium dioxide (TiO_2), poly(acrylic acid), and polyvinylpyrrolidone. Produced coating demonstrated impressive growth inhibition rates of 97% against *Staphylococcus aureus* and 96% against *Escherichia coli*. A mechanism for photoelectronic antibacterial and antifogging properties was proposed, highlighting positive photocatalytic and electrical characteristics of PEI-TiO_2.

Research contributes valuable knowledge for the design of multifunctional bioinspired surface materials that can effectively provide antifogging and antibacterial solutions for applications in both medical and everyday contexts. Biomimetic materials have huge scope in emerging fields of the research [13]. Inspired from morphology of the bone surfaces that have undergone osteoclast resorption, a Ca-doped TiP coating material characterized by nanofibrous web on to Ti using a straightforward two-step hydrothermal approach was developed [14]. This method aims to replicate the compositional and microstructural properties of natural bone. In vitro investigations indicate that Ca-TiP material, which features a synergetic combination of nanofibrous structure with the surface chemistry, can significantly promote osteogenic activity. This includes improved adhesion of cells, their distribution, and their proliferation as

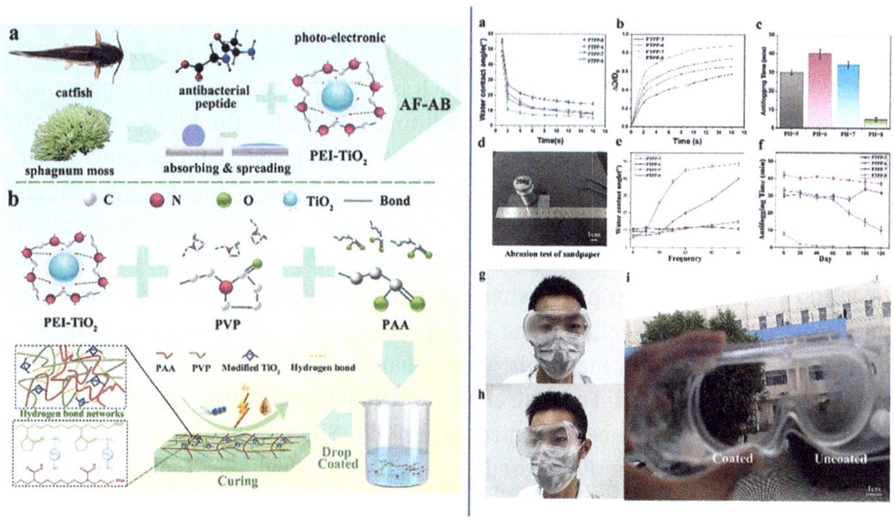

Fig. 5: (Left) (a) Catfish and sphagnum moss bionic prototypes and (b) schematic illustration of the preparation of the PTPP-coated substrates; (Right) (a) water contact angle evolution, (b) basal diameter changes of the water droplets, (c) antifogging durations of different PTPP coatings, (d) abrasion test of sandpaper, (e) water contact angle during the abrasion, (f) antifogging properties of different coatings, (g, h) wear antifogging tests of medical goggles, and (i) frost test of medical goggles. Copyright © 2024 American Chemical Society, reproduced with due permission from [12].

well as increased alkaline phosphatase activity along with enhanced expression of genes associated with bone formation, ultimately contributing to development of bone and enhancement of osseointegration quality. For clinical applications, the issue of infections related to biomedical devices is of critical importance. To reduce the risks associated with biomedical devices, improving surface properties (antibacterial and lubricating effectiveness) is viewed as a deliberate measure. For that, copolymer that possesses self-adhesive properties, lubricating features, and antibacterial properties was synthesized using free radical polymerization combined with a carbodiimide coupling reaction [15]. Antibacterial effectiveness of copolymer-modified surface system was suggestively enhanced against *S. aureus* and *E. coli*. Furthermore, in vitro assessments of the compatibility of blood as well as cells highlighted exceptional anticoagulant properties of the coating.

4.2 Advanced Coatings for Space Application

Mechanical components intended for future missions to Venus will be subjected to extreme conditions characterized by high temperature and pressure carbon dioxide environments. In that direction, surface of Inconel 718 (INC 718), a Ni-based alloy suitable for high-temperature applications, was enhanced through the application of advanced coatings using various deposition methods [16]. Mainly three commercially available coatings were investigated: Nedox PF-F™, Cera Tough™, and Balinit® DLC. These coatings were paired with PS400, which is a NASA-developed high temperature alloy, subjected to sliding tests under conditions simulating the high-pressure and high-temperature environment of Venus. The results indicated that changing the surfaces with PS400 led to a 25% reduction in friction; however, this modification also resulted in a significant increase in wear compared to traditional metal-on-metal contact. Coefficient of the friction is declined 38% for PS400 versus $TiMoS_2$ and by 43% for PS400 versus DLC tribo pairs, when compared with the friction observed in INC 718 versus INC 718 interactions. The aerospace sector is persistently investigating lightweight, high-strength materials, with a particular focus on carbon fiber reinforced polymer composites. For that, silane-functionalized GO was applied to surface of carbon fibers to enhance their properties [17]. Silane-functionalized graphene oxide was deposited on the PAN-based carbon fiber through electrophoretic deposition technique. At cryogenic temperatures, these improvements were recorded at 15.1%, 35.3%, and 13.5%, respectively.

Since cutting-edge-coating systems derived from polymer composite utilized in the spacecraft are significantly impacted by the severe conditions of the space environment including atomic oxygen, intense UV radiation, thermal cycling, and space wreckage. These factors lead to the development of cracks, particularly due to the pervasive effects of strong UV radiation, thereby considerably reducing the service life of these materials during prolonged missions. In response to this challenge, a novel UV-responsive coating system was established to facilitate in-orbit fault repair [18]. This coating comprises microcapsules with an internal shell that is rapidly degraded by an external layer of titanium dioxide on exposure to UV rays. The dual release mechanism of the UV-responsive microcapsules enhances the efficiency of agent release compared to conventional crack-repair systems. Furthermore, the detrimental effects of UV rays in space may be transformed to beneficial outcomes, indicating a promising application for this technology in aerospace coatings. Techniques such as ALD can create protective films that address several challenges faced by polymers in space [19]. Experiments involving atomic oxygen and vacuum ultraviolet were performed using laser-detonation source to generate hyperthermal oxygen atom and a D2 to provide vacuum ultraviolet light. The application of aluminum oxide ALD material for coating effectively shielded underlying materials from atomic oxygen damage, while incorporation of titanium oxide coating material further enhanced protection against VUV radiation. As we know that the utilization of UV-triggered self-healing mechanisms in individual microcapsules presents a promising approach to prolonging the lifespan of polymer-based aerospace coatings. For that, a robust self-healing system was developed utilizing single SiO_2 microcapsules, which operate through UV-triggered polymerization, aimed at possibly used for the coatings for in aerospace applications [20]. By leveraging resemblance in solubility between SiO_2 and active healing agents, epoxy resin as well as photoinitiator were capsulated within single microcapsule of SiO_2.

This performance is ascribed to a high concentration of agents and a carefully calibrated stoichiometric proportion of photoinitiator and epoxy. Crucially, the healing mechanism assisted by UV-triggered polymerization is not adversely affected by oxygen, thereby enhancing the feasibility of integrating these single SiO_2 microcapsules into spacecraft coatings, facilitating self-healing capabilities in environments characterized by abundant UV radiation and oxygen.

Future space missions are expected to function in increasingly challenging environments, including low-perihelion orbits as well as the Jovian magnetosphere. The investigation necessitates a careful choice of optically important materials and the components that can withstand various environmental factors. To simulate space conditions on the Earth, ion accelerators are employed. Systematic investigations have been conducted on effects of helium particles from solar wind striking to the thin film of Au, with absorbed doses [21]. The findings are analyzed in terms of the consistency and limits of laboratory experiments, providing crucial insights for the procedures involved in space qualification.

Also, within a spacecraft, the environmental conditions of temperature and humidity that are conducive to human habitation also inadvertently foster an optimal setting for the growth of bacteria and fungi. This situation poses significant risks to human health and can compromise the reliable operation of onboard equipment. To mitigate these concerns, wear-resistant as well as antimicrobial films were produced aimed at coating critical internal components of spacecraft [22] as shown in Fig. 6. Since, Ag and Cu are widely researched antimicrobial agents; therefore, researchers study focused on examining antibacterial efficacy of the coating of amorphous carbon that were doped with Ag, a combination of Ag and Cu, or Ag clusters. Durability of produced antimicrobial coating systems that is significantly affected by the diffusion of metals within coating material was assessed. The findings indicate that while silver-doped coatings exhibit exceptionally high antimicrobial activity, coatings doped with silver clusters demonstrate comparable effectiveness and enhanced longevity, despite having a lower overall concentration of silver.

Polyimides are extensively used for thermal blankets that shield outer surfaces of spacecraft due to the durability of polyimide in space and favorable thermal and optical characteristics. Nonetheless, these materials are vulnerable to erosion caused by atomic oxygen, which poses a significant risk in low Earth orbit as well as to electrical charging. A study illustrates that the application of 100 nm of SnO_2 via liquid-phase deposition forms a robust protective layer on Kapton polyimide, effectively mitigating atomic oxygen-induced erosion of surface and minimizing electrical charging [23]. SnO_2 layer, once deposited, does not significantly alter original characteristics of polyimides and successfully prevents electrostatic discharge. The benefits of deposited oxides by liquid phase, particularly their nonreliance on line-of-sight application, are well-recognized, and this research expands those benefits to include coatings which effectively recline electrostatic discharge during maintaining guard against atomic oxygen erosion. In another work, cellulose nanocrystals were used as bio-derived nano-

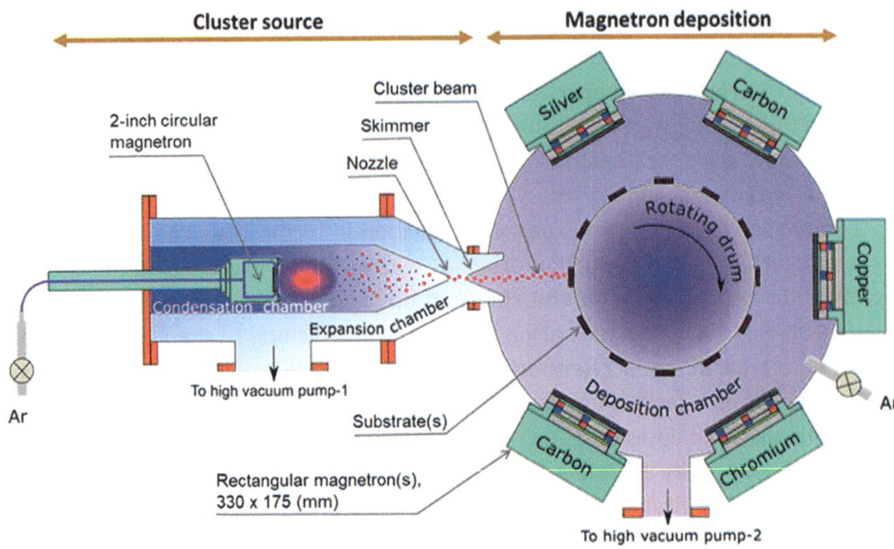

Fig. 6: Schematic of the vacuum system used for the deposition of carbon thin films. Copyright © 2022 American Chemical Society, reproduced with due permission from [22].

additives for defensive polydimethylsiloxane (PDMS)-coating material for space applications in order to advance thermal as well as optical aspects. Cross-linked PDMS material was then examined for mechanical features, thermal steadiness, and optical attributes post-photoaging. The outcomes showed that cellulose nanocrystals were found to be meaningfully taking part to advance thermal steadiness of produced coating material up to the temperature of 140 °C. Also, lightweight, durable, and adaptable polymer-based coating structures that are compactly foldable for transport and subsequently deployed in the space, experience repetitive flexing prior to launch and significant degradation due to experience to atomic oxygen. While numerous atomic oxygen-resistant coatings are being created, finding an advanced coating system which fulfills the essential criteria for both mechanical strength and long-standing guard for membranes remains a challenge. For that, a coating was developed that combines mechanical strength with prolonged space durability, utilizing inorganic polymer [24]. Initially, a coating consisting of polymer/silica bicontinuous phase was applied on ground, demonstrating remarkable elasticity along with superior resistance for abrasion. Subsequently, coating undergoes in situ self-evolution process when exposed to synergistic effects of atomic oxygen and UV rays, resulting in a dense, crack-free silica layer with exceptional durability.

Another work involved the synthesis of UV-curable cycloaliphatic polyesters [25]. It was observed that the T_g, tensile properties, as well as crosslinked polyester modulus increased thoroughly upon loading level of spiroglycol. In comparison to the aromatic controls, cycloaliphatic coatings demonstrated similar resistance for corrosion and weatherability along with excellent impact resistance. The results indicate that cycloaliphatic coatings present a viable alternative to aromatic-based systems for exterior durable clear coat applications.

4.3 Underwater and Marine Applications

Marine coatings are specifically designed protective layers utilized on a range of marine structures and assets to protect them from corrosion, fouling, and other types of deterioration resulting from exposure to seawater and environmental conditions. The graphite-like coating (GLC) is among the attention-seeker capable protective material system for the marine uses. For that, a multifunctional (Cr, Cu)-GLC-coating material was developed with collective tribocorrosion and antifouling features using sputtering technique [26]. The findings from the experiments suggest that obtained coatings transitioned from a compact structure to a more loosely arranged columnar structure as levels of Cr and Cu doping increased. Concurrently, hardness of coating exhibited a gradual decline. Furthermore, algae adhesion assessment demonstrates that density of algae on to surface of the (Cr, Cu)-GLC coating diminishes from approximately 565–70/mm^2 with rise in doping levels. Due to the prevalence of oil spill incidents in marine environments, there is a significant demand for stable superoleophobic coatings that can function effectively in seawater. Drawing from the remarkable mechanical characteristics of the seashell nacre as well as naturally occurring superoleophobic surfaces, a composite coating using a conventional assembly technique, utilizing high surface energy building blocks was developed [27]. The "bricks-and-mortar" architecture of nacre has been theoretically mirrored in the synthesized film, imparting exceptional mechanical strength and significant abrasion resistance. Furthermore, the resulting film demonstrates steady underwater superoleophobicity, minimal oil adhesion, along with remarkable environmental durability.

A novel strategy was used to facilitate the in situ creation of an underwater superoleophobic surface using sea water [28]. This process begins with the application of coatings made from self-polishing polymer combined with polymer-grafted SiO_2 NPs. Surfaces of these coatings demonstrate the ability to regenerate in artificial seawater by hydrolysis of outer layer of self-polishing material, followed by the dissolution. Notably, grafted copolymer chains can undergo a transformation into hydrophilic structures due to seawater-induced hydrolysis that enhances oil repulsion and provides surface with remarkable resistance to protein adsorption. Biofouling presents a significant challenge to the marine industry, as the accumulation of microcontaminants and macrocontaminants may lead to increased fuel ingesting along with potential harm to the hulls of ship. Furthermore, adjusting the zwitterionic hydrogel-derived coating material to enhance the antisediment aspects, aiming to attain the performance levels comparable to those of commercial coating materials those are regarded as benchmark in the antifouling technology was developed [29]. Following a week of sediment exposure and 2 weeks of diatom co-culturing, the optimized zwitterionic hydrogel-derived coating material demonstrated sustained antifouling capabilities, with minimal diatom presence on its surface. Additionally, large-scale samples exhibited antifouling performance akin to that of biocidal coatings in Atlantic Ocean over a duration of 1.5 months. In another study, an innovative and effective antifouling coating was developed that employs a synergistic approach, integrating mechanisms of interaction inhibition, fouling repulsion, and antifouling characteristics [30]. Experimental outcomes indicated that this coating effectively prevents adhesion of various protein-containing materials, various bacteria, along with *Navicula* diatoms.

Formation of biofilms on material surfaces poses significant challenges across various industries. For that, fluorinated carbon nanotubes were prepared using the one-step fluorination process to create a superamphiphobic materials for coating through straightforward spray application [31]. The coating materials, characterized by its extremely low surface free energy as well as steady micro/nanoframework, demonstrated remarkable and sustained antibiofouling capabilities. Substances such as milk, tea, bacterial solutions, and BSA exhibited significant bouncing behavior on this surface without any wetting in air. The persistent presence of an underwater layer on superamphiphobic layer of coating was noted, enabling it to effectively resist the attachment of BSA and bacteria, achieving efficiencies of 97.5% and 98.2%, respectively. Consequently, produced coating material shows substantial potential for applications in various marine uses, in clinical and biomedical materials, protecting clothing, as well as chemical shielding. In another work, charged organohydrogels were developed by amphiphilic and zwitterionic heteronetworks, synthesized through a straightforward emulsion polymerization process [32]. The charged organohydrogels benefit from a synergistic interplay of covalent bonds, electrostatic forces, and hydrogen bonding, which confer exceptional mechanical properties such as stretchability, compressibility, and recoverability. Due to spatial constraints imposed by the hydrophilic and oleophilic chains within the interpenetrating heteronetworks, charged organohydrogels demonstrate significantly low swelling behavior in both aqueous and oily environments, making them suitable for long-lasting applications without compromising their inherent performance. Regarding the correlation between structure and antifouling efficacy, elastic surfaces that integrate amphiphilic and zwitterionic properties demonstrate notable resistance to fouling from organisms along with potential for recyclability. The COHGs may thus represent a promising avenue for development of high-performance and environmentally friendly antifouling materials.

4.4 Nanoengineered Coatings and Thin Films

Nanocoatings consist of ultra-thin surface layers that can be composed of various materials, meticulously designed at the molecular or atomic scale. These coatings generally involve substances with structural dimensions on the order of nanometers. In that direction, a Janus-like configuration is observed for W-Mo films, characterized by a distinct separation of the two metals [33]. These thin width films were fabricated on the glass and silicon surfaces through co-sputtering method. Two series of films, with thicknesses of 300 nm and 1 μm, are produced under identical conditions of pressure and tungsten target current, while varying the Mo target currents between 10 and 200 mA. This dual-component architecture is consistently observed across all columns in developed film, confirming about the W-Mo films indeed showed Janus-like framework. Likewise, conformal growth of thin GAP coating layers was prepared using plasma-enhanced atomic layer deposition (PE-ALD) method [34]. GaP layers were deposited onto silicon wafers featuring a structured surface characterized by silicon microcolumns and black silicon. Consequently, this deposition technique is applicable for the conformal deposition of thin crystalline GaP layers on structured silicon surfaces that possess a high aspect ratio.

Surface patterning represents a compelling strategy for altering the surfaces of biomaterials, thereby influencing cellular activities and improving the efficacy of medical implants without resorting to conventional chemical modifications such as the incorporation of growth factors, antibiotics, or pharmaceuticals. Various nano-to-micron-scale patterns have been developed on thermoplastic and thermoset polymer coatings applied to bioresorbable magnesium substrates, aimed at regulating cellular responses and material degradation in vascular applications [35]. The technique of capillary force lithography was enhanced and combined with spray coating to create precisely aligned nano-to-micron patterns on thermoplastic poly(lactic-*co*-glycolic acid) (PLGA) and thermoset poly(glycerol sebacate) (PGS) coatings applied to magnesium substrates. The polymer coatings of PLGA and PGS, featuring nano-to-micron patterns on magnesium, were meticulously characterized, and their influence on cell adhesion and morphology was examined through direct in vitro culture with human umbilical vein endothelial cells. The PLGA coatings demonstrated a greater reduction in the degradation of magnesium substrates compared to the coatings observed during direct culture with cells in vitro. Morphology of the coating is shown in Fig. 7.

Furthermore, a bio-inspired graphene–epoxy coating material incorporating graphene as filler is developed using spraying technique [36]. Polydopamine was utilized to enhance the dispersion and rectify structural defects in graphene through π–π interactions, while also facilitating strong adhesion between dense layers of graphene and epoxy layers, thereby creating "interlock" structures that ensure the integrity of the coating system. Electrochemical assessments demonstrated that bio-inspired composite coating exhibited a significant increase in coating resistance. This bio-inspired approach presents an innovative method for the development of high-performance material for coatings with enhanced features.

Traditional disposable personal protective equipment primarily relies on mechanical filtration to obstruct pathogenic bacteria, which leaves a significant risk of contamination and transmission. Engineered polyphenol coatings, inspired by phenolic-enabled nanotechnology, utilizing aldehydes and metal interactions, referred to

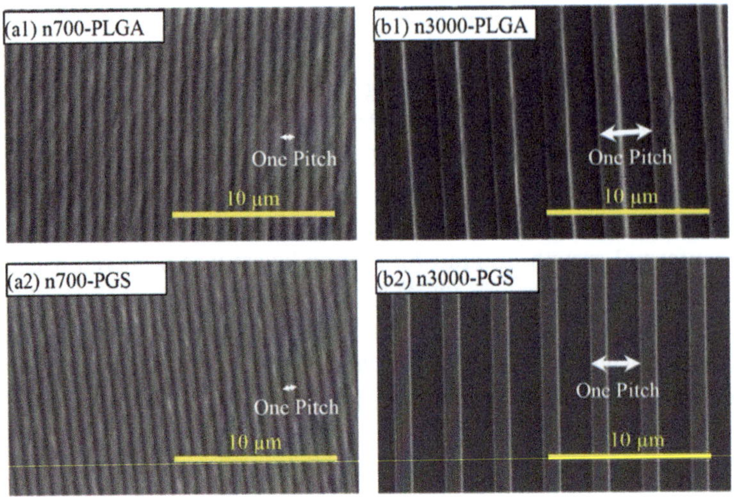

Fig. 7: Surface morphologies of nano-to-micron-patterned polymer coatings on Mg substrates. Copyright © 2020 American Chemical Society, reproduced with due permission from [35].

Fe-doped engineered polyphenols, were developed [37]. This approach aims to achieve effective photocatalysis and self-Fenton antibacterial activity. Investigations into the mechanism indicate that Fe-doped engineered polyphenols eliminate bacteria via direct reactive oxygen species attacks and gene regulation. Additionally, Fe-doped aromatic aldehyde has been successfully developed as an optimal antibacterial coating for various fabrics achieving over 93% bacterial reduction after five usage cycles. Nanoparticles coated with poly(ethylene oxide) (PEO) brushes, which consist of highly cross-linked copolymers having radii of approximately 30 nm were prepared [38]. In this setup, biodegradation was the sole mechanism for coating breakdown. Enrichment cultures capable of degrading PEO provided homopolymer or brush NPs as exclusive source, and the production of protein and carbon dioxide was observed to assess biological conversion. The biodegradation of the brush led to particle aggregation, highlighting the necessity of understanding biologically facilitated conversions of NPs coatings to better comprehend fate and transportation of NPs in environmental contexts.

4.5 Coatings and Surface Engineering for Sustainable Tribological Applications

The application of molybdenum disulfide (MoS_2) coatings through PVD and mechanical burnishing were extensively studied for use as solid lubricants in space environments; however, these methods are limited by line-of-sight issues when applied to complex geometries. The electrodeposited $Mo_xS_yO_z$ and $Mo_xS_yO_z$/Ni coatings films were developed using ammonium tetrathiomolybdate [39]. These coatings exhibit exceptional properties including lower coefficients of friction along with higher wear resistance. The characterization of morphology of these coating materials reveals the presence of amorphous microstructures formed by elevated oxygen and cracks upon drying. Furthermore, transition metal chalcogenides characterized by a lamellar structure are recognized for their applications in tribology, albeit primarily in vacuum conditions due to their susceptibility to degradation when exposed to oxygen and moisture. For that, a customized WSex coating that exhibits low friction and minimal wear rates in ambient air was developed [40]. Comprehensive analysis of nanostructure of film and its composition was performed using advanced TEM, revealing that surface layer consists of WSe_2 implanted within a W-rich amorphous matrix, designated as a-W(Se). Post-friction testing indicated an enhancement in the crystalline order along with alignment of WSe_2 in direction of sliding within interfacial region.

Since feasibility of utilizing a fluorinated boron nitride nanosheets (F-BNNSs)-based lubricant material for hot-forming processes involving Ti alloys. In that direction, F-BNNSs were synthesized via wet ball milling, combining hexagonal boron nitride (h-BN) with NH_4F, and resultant coating material was applied using a spray technique [41]. Tribological performance of these coatings was assessed on friction pairs composed of pure titanium (TA1) and silicon nitride (Si_3N_4) at a temperature of 800 °C. Findings indicated that F-BNNSs coatings demonstrated remarkable thermal stability, with the F-BNNSs3 variant achieving most favorable tribological outcomes. Consequently, F-BNNSs coating presents significant potential for application for hot-forming of Ti alloys.

Furthermore, continuous development of CrN coating material for up to 24 h has been successfully demonstrated, resulting in the formation of ultrathick layers [42]. Results from scratch and friction tests reveal a marked enhancement in load-bearing capability of the coating as thickness of CrN layers increases. Notably, during scratch test, ultrathick CrN coating material withstood a load of 180 N without failure, with cohesive failure mechanisms such as wedge spallation and cracking being predominant. The exceptional load-bearing capacity and superior tribological properties are due to relaxation mechanism and the reduced contact pressure associated with the continuous growth of the coating. Another study suggests to develop a graphene-reinforced Cr_3C_2NiCr composite coating through process of shroud plasma spraying, ensuring simultaneous exfoliation of graphene [43]. Mechanical properties of coating, which contains 2% reinforcement, exhibited enhancements of 70.4% in hardness, 31.8% in elastic modulus, and 61.5% in fracture toughness. These improvements are primarily due to the uniform distribution of graphene, as well as mechanisms such as crack bridging and crack arrest that are promoted by presence of exfoliated graphene in composite matrix. Additionally, wear volume loss, wear rate, and coefficient of friction for the coating were reduced by 55.7%, 55.7%, and 78.7%, respectively. These results underscore potential of exfoliated graphene to enhance performance of coatings across a range of manufacturing uses.

Titanium alloys are widely utilized as materials for implants in artificial joints due to remarkable biocompatibility, superior mechanical features, higher strength-to-weight ratio, and commendable corrosion resistance. The chemical self-assembly of innovative 2D diamond coatings on titanium alloys, in conjunction with natural SF, results in significantly enhanced friction as well as wear properties [44]. Linear-reciprocating sliding tests reveal that the friction coefficient and wear rate of the diamond nanosheet coating, when lubricated with SF, are reduced by 54% and 98%, respectively, in comparison to the uncoated titanium alloy that is lubricated with water. The lubrication mechanism of this innovative system was clarified through an extensive examination of the microstructural and chemical changes that occur. The lubricating mechanism in PA-DN/SF system is shown in Fig. 8.

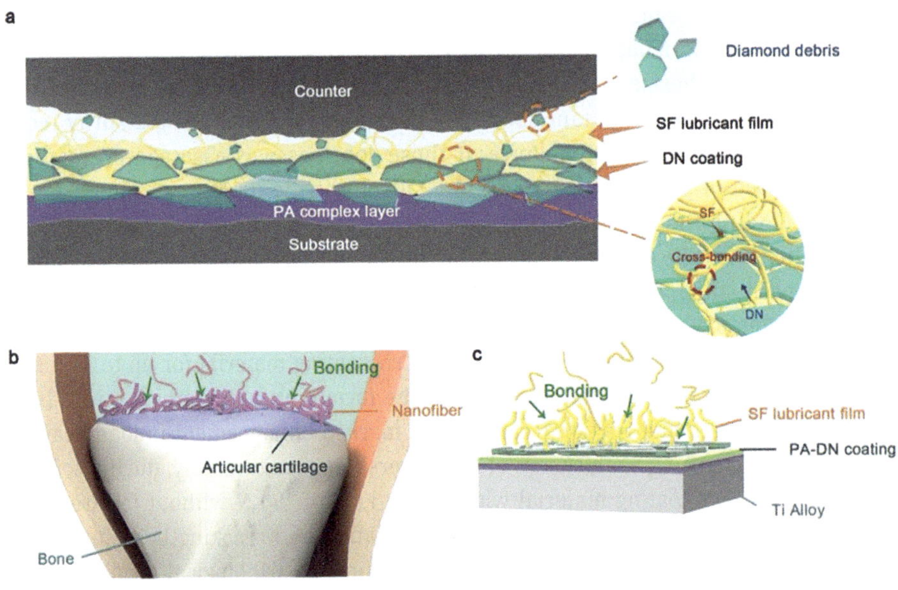

Fig. 8: (a) Lubricating mechanism in PA-DN/SF system, (b) mechanism between cartilage and synovial fluid, and (c) construction of PA-DN/SF system on Ti alloy. Copyright © 2022 American Chemical Society, reused with due permission from [44].

The friction interfaces found in liquid-embedded composite lubrication coatings generally comprise a mixture of discontinuous fluid films and rough solid contact surfaces, which promote effective shearing and prolong wear life. Nevertheless, attaining optimal performance in mixed lubrication poses difficulties due to the fundamental conflict between the improved mobility of the liquid lubricant and the heightened mechanical strength of the solid matrix. Recent studies have focused on the integration of reinforcing fillers to develop multicomponent, multiphase composites capable of addressing this challenge. In this regard, a modified attapulgite (APT) has been introduced, demonstrating robust biphasic wettability, which is achieved through an oil decompressive osmosis treatment applied to APT nanocontainers that are modified with long nonpolar alkyl chains [45]. This revised APT facilitates the control of the size, distribution, and movement of lubricant droplets by creating a Pickering emulsion, while concurrently improving the solid-phase matrix via dispersion strengthening. Moreover, the incorporation of APT encourages the formation of a solid tribofilm during friction, which exhibits an enhanced oil adsorption capacity, as validated by first-principle calculations employing density functional theory.

4.6 Plasma Surface Coating

A significant rise in worldwide atmospheric temperatures was experienced in the past few years. The insufficient provision of protective measures against sun exposure during working hours has led to an increase in sunburn incidents among employees. It is crucial to foster productive work environments while ensuring safety and health are not compromised. In that direction, the application of low-temperature plasma (LTP) technology to enhance surface characteristics of fabrics, enabling them to reflect IR radiation from sunlight was analyzed [46]. The LTP method is particularly suitable for the temperature-sensitive materials attributed to their operational range of 30 °C. Commercially available fabrics and electrospun PET surfaces were treated using tetraethoxy orthosilicate (TEOS) plasma. This treatment facilitates the deposition of a reactive plasma-polymerized silane nanolayer on surfaces of such materials. SEM and TEM analyses indicated that the nanoparticles formed were within the size range of 100–200 nm. Collectively, these findings underscore a promising potential of plasma processing in advancement of infrared reflective coating material.

Furthermore, a novel atmospheric plasma posttreatment approach was established for surface coating of 3D structures fabricated through 3D printing [47]. Study examined about various effects of nonequilibrium atmospheric plasma for various chemical as well as physical characteristics, comparing these effects to those of conventional posttreatment using UV light. Additionally, thermosets that did not undergo any posttreatment were included in the coating analysis for comparison. The findings indicate that ultraviolet posttreatment may facilitate accomplishment of curing in printed material when full curing cannot achieve while the ongoing process of printing. In contrast, plasma posttreatment cannot enhance curing of material; instead, it leads to regeneration of aldehyde functionalities. Notably, the noncoated thermosets also exhibit significant antioxidant properties, which are inferred from their chemical structures. Plasma polymerization has gained recognition as a promising method for surface modification due to its numerous advantages over traditional techniques, such as user-friendliness and the ability to modify a wide range of substrates. An alternative plasma-based approach for creating coatings with specific functionalities was developed [48]. Initially, hexamethyldisiloxane plasma polymerization is conducted in a medium-pressure dielectric barrier discharge system to deposit coatings resembling PDMS. Subsequently, the resulting coating is subjected to air plasma treatment within a similar dielectric barrier discharge configuration to incorporate silanol groups on surface. Silanol groups serve as sites for the grafting of (3-aminopropyl)triethoxysilane and (3-bromopropyl)trichlorosilane in the final step, allowing for the selective introduction of amino or bromo functionalities, respectively. Furthermore, the coatings were successfully synthesized on three distinct substrates demonstrating a broad applicability of proposed method.

The issue of infections was addressed by creating innovative composite coatings of HAp and gallium liquid metal (GaLM) through the atmospheric plasma spray (APS) technique [49] as shown in Fig. 9. A mixture containing 5 wt.% Ga was incorporated into commercially available HAp powder using an orbital shaker; subsequently, HAp-Ga particle feedstock was applied to Ti_6Al_4V substrates via APS method. GaLM effectively filled pores of HAp coating, particularly in areas of voids and cracks, thereby mitigating potential failures at these sites. The surface wettability transitioned from hydrophobic in HAp coating to hydrophilic in HAp-Ga composite coating. Furthermore, HAp-Ga coating demonstrated remarkable antibacterial efficacy against both initial bacterial adhesion and established biofilms formed by methicillin-resistant *S. aureus* and *P. aeruginosa*. The investigation indicates that GaLM enhances antibacterial properties of HAp-based coatings while preserving advantageous characteristics of traditional HAp coatings. Consequently, HAp-Ga APS coating emerges as a promising alternative for the antibacterial applications.

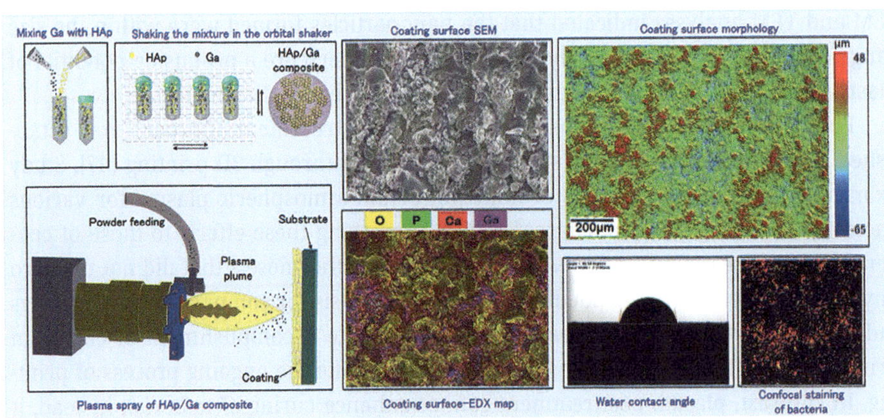

Fig. 9: Schematic of powder preparation and plasma spray coating along with morphology and confocal staining of bacteria. Copyright © 2022 American Chemical Society, reproduced with permission from [49].

The implant-associated infections represent a significant challenge in the context of total joint prostheses, with incidence rates reaching up to 3% during surgical procedures. For that, HAp was doped with 5% Zn^{2+} to mitigate bacterial proliferation, while an additional 5% Sr^{2+} was incorporated as a secondary dopant to alleviate cytotoxicity associated with Zn^{2+} ions [50]. Nanosized HAp powder was produced via a hydrothermal synthesis technique and subsequently subjected to heat treatment at 600 °C for a duration of 4 h. This treated powder was then applied to a Ti alloy substrate (Ti-6Al-4V) using plasma spraying techniques. The mechanical stability was assessed through adhesive bond testing, revealing an increase in strength. In parallel with mechanical enhancements, both antibacterial efficacy and biological functionality exhibited significant improvements. Notably, release of Zn^{2+} ions from coating was found to be influenced by the presence of Ca^{2+}, P, and Sr^{2+} ions, which in turn were dependent on the heat treatment temperatures.

Another study seeks to enhance interfacial adhesion between HAp particles and polylactide (PLA) to improve mechanical properties and biocompatibility of HAp/PLA-based bone implants [51]. To achieve this, a one-step surface functionalization of HAp through plasma polymerization is employed. Utilizing acetylene plasma chemistry, hydrophobic characteristics of HAp particles are precisely adjusted before their incorporation into a PLA matrix via an extrusion technique. Cytotoxicity of modified HAp particles is evaluated using WST-1 proliferation assay and lactate dehydrogenase release, revealing that cell viability consistently exceeded the 70% threshold, with negligible changes in lactate dehydrogenase levels, indicating favorable biocompatibility. Contact angle measurements, along with morphological and rheological assessments, demonstrate that lower plasma power results in more hydrophobic surfaces and an improved HAp/PLA interface. Dynamic mechanical analysis indicates that the storage modulus at 37 °C for composites containing functionalized particles increased by 1.5 times compared to those with unmodified particles.

5 Conclusion

Technological and scientific developments pave the way for new advances in field of smart surfaces and smart coating as it affects many areas of life. While conventional coatings are mostly used for physical protection and decorative purposes, a variety of functional smart coatings have emerged with these technological and scientific developments. The term “smart coatings” refers to the concept of a coating that can dynamically change its properties in response to a stimulus in environment. These external stimuli can be in the form of a physical interaction such as mechanical force, temperature, pressure, magnetic field, electric field, or electromagnetic radiation or changes in ambient conditions such as pH or chemical composition [52].

6 Future Trends and Developments

The future of intelligent surfaces and smart coatings is rooted in the creation of more resilient, eco-friendly, and adaptable formulations. Progress in nanotechnology and materials science is facilitating the development of coatings that integrate multiple smart features, thereby presenting new opportunities for applications such as self-cleaning, anti-icing, and water purification. Innovations in bio-based and recyclable materials for intelligent surfaces and smart coatings hold significant promise for minimizing environmental impact and promoting sustainability across various sectors. Continuous research and development in this area indicate a promising future for intelligent surfaces and smart coatings, with potential applications extending into novel and unforeseen fields. As far as future development of smart surfaces and smart coating materials is concerned, the recycling of polymers plays an important aspect to think beyond of the single-use materials for circular economy concern [53]. Recycled materials should be encouraged to use for nonsensitive types of coatings. Apart from this 3D and 4D printing along with machine learning should also be incorporated in the development of smart surfaces and smart coatings.

References

[1] Anonymous. The Open University. 2018 [cited 2025 Jun 21]. Thermal spraying (Hardfacing). Available from: https://www.open.edu/openlearn/science-maths-technology/engineering-technology/manupedia/thermal-spraying-hardfacing

[2] Becker M Fraunhofer Institute for Manufacturing Technology and Advanced Materials IFAM. 2025 [cited 2025 Jun 21]. Surface Technology. Available from: https://www.ifam.fraunhofer.de/en/technologies/wet-chemical-surface-treatment-for-better-bonds-and-coatings.html

[3] Khitab A, Abdin ZU, Ahmed I, Karim T Thermal insulation of buildings through classical materials and nanomaterials. In: Recent Advances in Nano-Tailored Multi-Functional Cementitious Composites. Elsevier; 2022. p. 277–303.

[4] Anonymous. A & A Coatings. [cited 2025 Jun 21]. 5 Different Types of Wear Resistant Coatings. Available from: https://www.thermalspray.com/5-different-types-of-wear-resistant-coatings/

[5] Ribeiro AM, Amaral C, Veiga F, Figueiras A Polymeric micelles as a versatile tool in oral chemotherapy. In: Design and Development of New Nanocarriers. Elsevier; 2018. p. 293–329.

[6] Anonymous. Hydromer. 2024 [cited 2025 Jun 21]. Differences Between Hydrophilic and Hydrophobic Coatings. Available from: https://hydromer.com/sb/differences-between-hydrophilic-and-hydrophobic-coatings/

[7] Cheng Q, Chen J, Cai W, Yu X, Wan C, Wang Y, et al. Biomimetic Colored Coating toward Robust Display under Hostile Conditions. ACS Appl Mater Interfaces. 2024 Sep 11;16(36):48448–56.

[8] Zhang Y, Tian J, Zhong J, Shi X Thin Nacre-Biomimetic Coating with Super-Anticorrosion Performance. ACS Nano. 2018 Oct 23;12(10):10189–200.

[9] Yang Y, Wang H, Yan FY, Qi Y, Lai YK, Zeng DM, et al. Bioinspired Porous Octacalcium Phosphate/Silk Fibroin Composite Coating Materials Prepared by Electrochemical Deposition. ACS Appl Mater Interfaces. 2015 Mar 18;7(10):5634–42.

[10] Li S, An Y, Zhao X, Zhou H, Chen J, Hou G Bioinspired Smart Coating with Superior Tribological Performance. ACS Appl Mater Interfaces. 2017 May 24;9(20):16745–49.
[11] Xiang T, Zheng S, Zhang M, Sadig HR, Li C Bioinspired Slippery Zinc Phosphate Coating for Sustainable Corrosion Protection. ACS Sustain Chem Eng. 2018 Aug 6;6(8):10960–68.
[12] Gao H, Xing Z, Liu J, Chen X, Zhou N, Zheng Y, et al. Bioinspired Photoelectronic Synergy Coating with Antifogging and Antibacterial Properties. Langmuir. 2024 May 21;40(20):10589–99.
[13] Singh A, Kumari K, Kundu PP Synthesis of Biomimetic Hydroxyapatite from Natural Sources for Bone Tissue Engineering. In: Encyclopedia of Green Materials. Singapore: Springer Nature Singapore; 2023. p. 1–14.
[14] Cai B, Tan P, Jiang N, Guo Z, Ay B, Li S, et al. Bioinspired Fabrication of Calcium-Doped TiP Coating with Nanofibrous Microstructure to Accelerate Osseointegration. Bioconjug Chem. 2020 Jun 17;31 (6):1641–50.
[15] Jia Y, Zhao Y, Zhang H Bioinspired Self-Adhesive Multifunctional Lubricated Coating for Biomedical Implant Applications. ACS Appl Bio Mater. 2024 Jul 15;7(7):4307–22.
[16] Tsigkis V, Bashandeh K, Lan P, Polycarpou AA Tribological Investigation of Advanced Coatings Subjected to Venusian Environment of 2.4 MPa CO_2 Pressure and 462 °C. Surf Coat Technol. 2022 Jul;441:128550.
[17] M M, Roy KER, MA M, Hassan SBA, Masihadas A. Effect of EPD Coated Silanized Graphene Oxide on Carbon Fiber Reinforced Plastic: An Emphasis on Mechanical Properties at Cryogenic Temperatures. Surf Coat Technol. 2022 Dec;451:129043.
[18] Zhu Y, Cao K, Chen M, Wu L Synthesis of UV-Responsive Self-Healing Microcapsules and Their Potential Application in Aerospace Coatings. ACS Appl Mater Interfaces. 2019 Sep 11;11(36):33314–22.
[19] Minton TK, Wu B, Zhang J, Lindholm NF, Abdulagatov AI, O'Patchen J, et al. Protecting Polymers in Space with Atomic Layer Deposition Coatings. ACS Appl Mater Interfaces. 2010 Sep 22;2(9):2515–20.
[20] Guo W, Jia Y, Tian K, Xu Z, Jiao J, Li R, et al. UV-Triggered Self-Healing of a Single Robust SiO_2 Microcapsule Based on Cationic Polymerization for Potential Application in Aerospace Coatings. ACS Appl Mater Interfaces. 2016 Aug 17;8(32):21046–54.
[21] Pelizzo MG, Corso AJ, Tessarolo E, Böttger R, Hübner R, Napolitani E, et al. Morphological and Functional Modifications of Optical Thin Films for Space Applications Irradiated with Low-Energy Helium Ions. ACS Appl Mater Interfaces. 2018 Oct 10;10(40):34781–91.
[22] Sanzone G, Field S, Lee D, Liu J, Ju P, Wang M, et al. Antimicrobial and Aging Properties of Ag-, Ag/Cu-, and Ag Cluster-Doped Amorphous Carbon Coatings Produced by Magnetron Sputtering for Space Applications. ACS Appl Mater Interfaces. 2022 Mar 2;14(8):10154–66.
[23] Gotlib-Vainstein K, Gouzman I, Girshevitz O, Bolker A, Atar N, Grossman E, et al. Liquid Phase Deposition of a Space-Durable, Antistatic SnO_2 Coating on Kapton. ACS Appl Mater Interfaces. 2015 Feb 18;7(6):3539–46.
[24] Wang D, Ma J, Li P, Fan L, Wu Y, Zhang Z, et al. Flexible Hard Coatings with Self-Evolution Behavior in a Low Earth Orbit Environment. ACS Appl Mater Interfaces. 2021 Sep 29;13(38):46003–14.
[25] Hammer TJ, Pugh C, Soucek MD. Ultraviolet-Curable Cycloaliphatic Polyesters Containing Spiroacetal Moieties for Application as Powder Coatings. ACS Appl Polym Mater. 2022 Apr 8;4(4):2294–305.
[26] Qiao Z, Yang R, Liu Y, Feng X, Huang J, Tian Y, et al. A Feasible Strategy to Construct WC-Co-Cr/cu Composite Coating with Good Tribo-corrosion Behavior and Anti-fouling Properties for Marine Applications. Wear. 2025 Mar;564–565:205738.
[27] Xu LP, Peng J, Liu Y, Wen Y, Zhang X, Jiang L, et al. Nacre-Inspired Design of Mechanical Stable Coating with Underwater Superoleophobicity. ACS Nano. 2013 Jun 25;7(6):5077–83.
[28] Wang D, Liu H, Yang J, Zhou S Seawater-Induced Healable Underwater Superoleophobic Antifouling Coatings. ACS Appl Mater Interfaces. 2019 Jan 9;11(1):1353–62.

[29] Song B, Zhang E, Shi Y, Wang W, Zhu H, Gallagher SJ, et al. Zwitterionic Hydrogel Coating with Antisediment Properties for Marine Antifouling Applications. ACS Appl Mater Interfaces. 2024 May 29;16(21):27908–16.

[30] Xie C, Guo H, Zhao W, Zhang L Environmentally Friendly Marine Antifouling Coating Based on a Synergistic Strategy. Langmuir. 2020 Mar 10;36(9):2396–402.

[31] Zhang D, Liu Z, Wu G, Yang Z, Cui Y, Li H, et al. Fluorinated Carbon Nanotube Superamphiphobic Coating for High-Efficiency and Long-Lasting Underwater Antibiofouling Surfaces. ACS Appl Bio Mater. 2021 Aug 16;4(8):6351–60.

[32] Zhang Z, Guo L, Hao J Emulsion-Based Organohydrogels with Switchable Wettability and Underwater Adhesion toward Durable and Ecofriendly Marine Antifouling Coatings. ACS Appl Polym Mater. 2021 Jun 11;3(6):3060–70.

[33] Boukhalfa H, Potin V, Martin N Structural and Electrical Properties of Nanocolumnar W-Mo Thin Films with a Janus-like Structure. Surf Coat Technol. 2022 Oct;448:128928.

[34] Uvarov A, Gudovskikh A, Baranov A, Maksimova A, Vyacheslavova E, Kirilenko D Conformal Growth of GaP on High Aspect Ratio Si Structured Surface via Plasma-enhanced Atomic Layer Deposition. Surf Coat Technol. 2024 Feb;477:130357.

[35] Jiang W, Zhang C, Tran L, Wang SG, Hakim AD, Liu H Engineering Nano-to-Micron-Patterned Polymer Coatings on Bioresorbable Magnesium for Controlling Human Endothelial Cell Adhesion and Morphology. ACS Biomater Sci Eng. 2020 Jul 13;6(7):3878–98.

[36] Ding J, Zhao H, Yu H. Bio-inspired Multifunctional Graphene–Epoxy Anticorrosion Coatings by Low-Defect Engineered Graphene. ACS Nano. 2022 Jan 25;16(1):710–20.

[37] Du T, Wang S, Feng J, Shen Y, Wang J, Zhang W Dual-Mechanism Tuned Engineered Polyphenols with Cascade Photocatalytic Self-Fenton Reaction for Sustainable Biocidal Coatings. Nano Lett. 2023 Oct 25;23(20):9563–70.

[38] Kirschling TL, Golas PL, Unrine JM, Matyjaszewski K, Gregory KB, V. LG, et al. Microbial Bioavailability of Covalently Bound Polymer Coatings on Model Engineered Nanomaterials. Environ Sci Technol. 2011 Jun 15;45(12):5253–59.

[39] Babuska TF, Reed CL, Banga D, Larson SR, Mings A, Curry JF, et al. Electrodeposited $Mo_X S_Y O_Z$/Ni Tribological Coatings. Langmuir. 2024 Mar 19;40(11):5776–84.

[40] Domínguez-Meister S, Conte M, Igartua A, Rojas TC, Sánchez-López JC. Self-Lubricity of WSe_X Nanocomposite Coatings. ACS Appl Mater Interfaces. 2015 Apr 22;7(15):7979–86.

[41] Ding S, Chang W, Wang W, Gong P, Wang C, Lv F, et al. Fluorinated Boron Nitride Nanosheets-Based Solid Lubricating Coating for Tribological Applications at Elevated Temperatures. ACS Appl Nano Mater. 2024 Sep 27;7(18):21871–81.

[42] Li Z, Wang Y, Cheng X, Zeng Z, Li J, Lu X, et al. Continuously Growing Ultrathick CrN Coating to Achieve High Load-Bearing Capacity and Good Tribological Property. ACS Appl Mater Interfaces. 2018 Jan 24;10(3):2965–75.

[43] Maurya SS, Grain AK, Pandey KK, Keshri AK Graphene Nanoplatelets for Reinforced Cermet Coatings with Improved Mechanical and Tribological Properties. ACS Appl Nano Mater. 2024 Aug 23;7(16):18811–23.

[44] Chen H, Cai T, Ruan X, Jiao C, Xia J, Wei X, et al. Outstanding Bio-Tribological Performance Induced by the Synergistic Effect of 2D Diamond Nanosheet Coating and Silk Fibroin. ACS Appl Mater Interfaces. 2022 Oct 26;14(42):48091–105.

[45] Li F, Chen L, Ma Y, Ji L, Lu Y, Wang A, et al. Microvesicle-Embedded Solid–liquid Composite Coating for the Tribological Behavior Regulation and Long-Acting Lubrication. ACS Appl Mater Interfaces. 2024 Jan 17;16(2):2774–87.

[46] Pillai RR, Appavoo K, Lungu CT, Behura SK, Thomas V. Plasma-Corona-Processed Nanostructured Coating for Thermoregulative Textiles. ACS Appl Polym Mater. 2024 Dec 13;6(23):14047–56.

[47] Liguori A, Xu H, Hazarika D, Hakkarainen M. Simple Non-Equilibrium Atmospheric Plasma Post-Treatment Strategy for Surface Coating of Digital Light Processed 3D-Printed Vanillin-Based Schiff-Base Thermosets. ACS Appl Polym Mater. 2023 Oct 13;5(10):8506–17.

[48] Egghe T, Ghobeira R, Esbah Tabaei PS, Morent R, Hoogenboom R, De Geyter N Silanization of Plasma-Activated Hexamethyldisiloxane-Based Plasma Polymers for Substrate-Independent Deposition of Coatings with Controlled Surface Chemistry. ACS Appl Mater Interfaces. 2022 Jan 26;14 (3):4620–36.

[49] Pham DQ, Gangadoo S, Berndt CC, Chapman J, Zhai J, Vasilev K, et al. Antibacterial Longevity of a Novel Gallium Liquid Metal/Hydroxyapatite Composite Coating Fabricated by Plasma Spray. ACS Appl Mater Interfaces. 2022 Apr 27;14(16):18974–88.

[50] Ullah I, Siddiqui MA, Liu H, Kolawole SK, Zhang J, Zhang S, et al. Mechanical, Biological, and Antibacterial Characteristics of Plasma-Sprayed (Sr, Zn) Substituted Hydroxyapatite Coating. ACS Biomater Sci Eng. 2020 Mar 9;6(3):1355–66.

[51] Carette X, Dhond L, Hemberg A, Thiry D, Mincheva R, Cailloux J, et al. Innovative One-Shot Paradigm to Tune Filler–Polymer Matrix Interface Properties by Plasma Polymer Coating in Osteosynthesis Applications. ACS Appl Bio Mater. 2021 Apr 19;4(4):3067–78.

[52] Rossi S, Simeoni M, Quaranta A Behavior of Chromogenic Pigments and Influence of Binder in Organic Smart Coatings. Dyes Pigm. 2021 Jan;184:108879.

[53] Singh A, Banerjee SL, Kumari K, Kundu PP Recent Innovations in Chemical Recycling of Polyethylene Terephthalate Waste: A Circular Economy Approach Toward Sustainability. In: Handbook of Solid Waste Management. Singapore: Springer Nature Singapore; 2022. p. 1149–76.

Parveen Kumar*, Devesh Kumar Pandey, Mansish Tomar, Gaurav Sharma, and Rajeev Sindhu

Emerging Technologies in Cementitious Material Design: Autonomous Healing and Durability Enhancement Mechanisms

Abstract: Cementitious materials provide the basis of most modern construction but are plagued with problems associated with durability, cracking and maintenance which could ultimately hinder the long-term performance and sustainability of construction structures. These problems can rarely be solved with conventional repair methods, and that is why new strategies for increasing the durability of these materials are required. The current innovational trends in various self-healing techniques including bacterial induced calcite precipitation, encapsulated healing agents as well as crystalline admixtures are evident solutions to the above challenges. These techniques are used to provide structure repair, seepage control, and extend the durability of cementitious structures up to several generations from the time of treatment without seeing or needing repairs again.

This review systematically discusses the most recent developments of self-healing systems with particular emphasis on their working principles, efficiency, and real-world usability. Self-healing strategies examined include bacterial autonomic systems that utilize microbial behaviour to fill the crack, encapsulation methods for using the crack to release the agent, and the use of nanotechnology for improvement in self-healing efficacy. This enables comparative analysis of these technologies side by favouring their duration enhancement capability environmental sustainability and scalability for industrial application aspects among others.

The results prove the generation shift of cementitious materials leading to self-healing and sustainable structures. However, practical difficulties including the stability of the healing agents and the cost-effectiveness of large-scale installation are complex issues of concern. It is therefore important to approach such issues through a multi-disciplinary research and development efforts in order to scale up the use of autonomous healing technologies in the construction industry. Potential future enhancements include the real-time monitoring of artificial intelligence and the Internet

*Corresponding author: **Parveen Kumar,** Skill Department of Automotive Studies, Shri Vishwakarma Skill University, Dudhola, Palwal 121102, Haryana, India, e-mail: parveenkumarme@gmail.com
Devesh Kumar Pandey, Mansish Tomar, Department of Civil Engineering, Rawal Institute of Engineering and Technology, Faridabad 121004, Haryana, India
Gaurav Sharma, Department of Mathematics, Rawal Institute of Engineering and Technology, Faridabad 121004, Haryana, India
Rajeev Sindhu, Department of BCA and B.Sc. (Computer Science), GNIOT Institute of Professional Studies, Knowledge Park-2, Greater Noida 201310, Uttar Pradesh, India

https://doi.org/10.1515/9783111515717-002

of Things and designing additional-agent structures using a combination of those powerful healing techniques in order to acquire better results.

Keywords: Cementitious materials, sustainability, self-healing, healing agents, nanotechnology

1 Introduction

1.1 Background and Importance of Cementitious Materials in Constructions

These include the cementitious products which include the concrete and mortar; the products that are backbone in the modern structure include building, bridges, and other important structures. They are used all over the world for a variety of civil engineering applications, greatly enhancing economic returns and social advancement [1, 2]. However, cementitious materials are inherently susceptible to problems such as cracking and deterioration in durability as a result of problems like shrinkage, thermal stresses, and environmental effects [3, 4]. Such issues require regular servicing and maintenance, as well as timely repairs, consequently raising the life cycle costs and the subsequent environmental impacts consequently [5].

Different studies show that microcracks in concrete are usually continuous and once developed, they lead to macro cracks affecting the overall strength and durability and permeability to water, chlorides, and sulphates [6, 7]. These accumulations consist of reinforcement corrosion, spalling, and reduction in the service life of the structure [8]. For example, De Belie [9] and Mohamed et al. [10] reported that cracking accounts for approximately 38% of failure in concrete structures internationally. These problems highlight the importance of dealing with the inadequacies of traditional cementitious materials and the necessity of the search for new possibilities to improve their resistance [11, 12].

1.2 The Need for Autonomous Healing Mechanism

The use of traditional repair techniques such as hand-applied patching and epoxy resin injection is costly, slow and does not have a significant impact on the affected areas particularly those that are difficult to access [11, 12]. Moreover, the said methods do not include treatments of internal microcracks which are influential for durability of concrete structures [13]. This is true especially because, as Khushnood et al. [14] and Xue et al. [15] have pointed out, relying on a reactive maintenance approach is financially and environmentally unfeasible especially when considered against the background of increasing infrastructural development needs all across the world [15, 16].

Autonomous self-healing cementitious materials are an exciting development that might ensure crack repair and durability of structures [17]. Autonomous mechanisms are derived from biological structures and processes where damage resets an inherent self-repair response [17–19]. Some current research works like the works done by Feng et al. [19] and Fang et al. [20] have revealed that self-healing technologies are efficient when it comes to sealing of microcracks, the reduction of permeability and the enhancement of service lives. These advancements are within the sustainability objectives because they cut out the frequency of repair interventions, hence conserving material use [21].

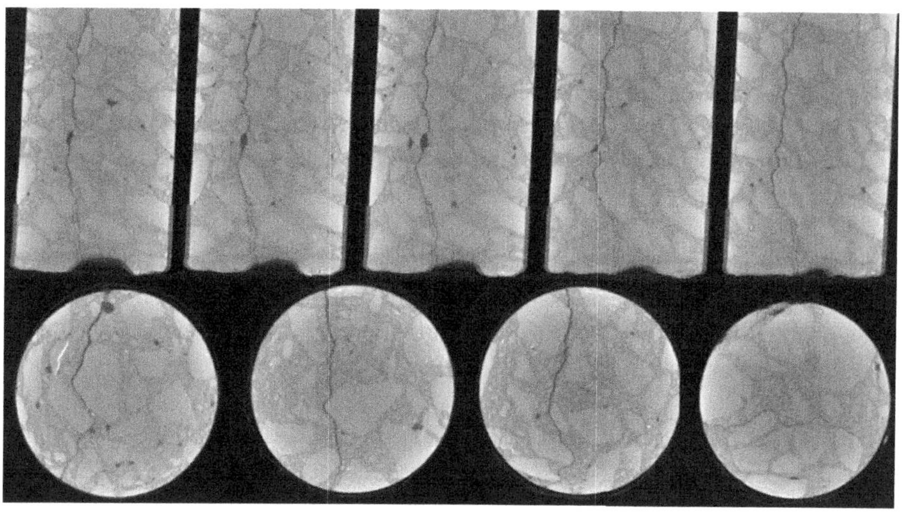

Fig. 1: Mechanisms of crack formation and propagation in cementitious materials (source: Wang et al. [20], adopted under the terms and conditions of the creative commons attribution (CC BY) license (https://creativecommons.org/licenses/by/4.0/)).

Figure 1 depicts the step-by-step process of crack formation and growth in cementitious materials with emphasis to the mechanisms that allow the penetration of water, ions and other adverse species. The figure idealizes how and why microcracks, which originate from shrinkage, thermal stresses, or mechanical loading deteriorate in size and jeopardize concrete structure. It also shows that crack development and durability concerns such as the corrosion of steel reinforcement and deterioration of materials are directly related [21, 22].

This visualization extends Tab. 1 by supporting the issues elaborated on previously, including cracking and environmental degradation. It brings focus on the need to solve such vulnerabilities because De Belie [9] and Mohamed et al. [10] have attributed about 40% of structural failure of cementitious materials to them. Furthermore, the figure highlights disadvantages of other conventional repair techniques such as epoxy injection which do not consider cracks within the structure [11]. Combined with Tab. 1, it supports the need for autonomous healing technologies to address the problem of durability [13, 15].

1.3 Objectives and Scope of the Review

The objective of this review is to present a critical appraisal of the current trends in technologies to self-heal cementitious materials. The key objectives are:

1. To compare different self-healing methods such as autogenous healing, MICP, encapsulation technique, and chemicals [21–24].
2. To evaluate the development and the progress in the area of improving durability with the help of self-healing technologies with the consideration of their efficiency and environmental friendliness [25, 26].
3. To define further directions of investigating the future trends and utilizing the – Application of hybrid systems, nanotechnology, and artificial intelligence in the field [27–30, 31].

The review applies perspectives from materials science, microbiology, and civil engineering disciplines to arrive at the findings. Through integrating results from seventy-three references, it intends to fill existing gaps and offer practical suggestions for researchers and stakeholders.

Tab. 1: Summary of challenges and opportunities in cementitious material design.

Challenge	Traditional solutions	Limitations	Emerging solutions	References
Cracking and durability issues	Manual repairs, epoxy injection	High cost, limited effectiveness	Self-healing technologies	[1, 10, 14]
Environmental segradation	Protective coatings	Requires frequent application	Autonomous healing agents	[6, 15, 20]
High life cycle costs	Reactive maintenance	Unsustainable	Sustainable materials	[8, 13, 17]

Table 1 offers a brief explanation of the major difficulties related to cementitious material design, the conventional strategies used to address these problems, the drawbacks related to the utilization of these strategies, and the new solutions in the form of innovative technologies. The table identifies three critical challenges: performance problems such as cracking and durability, problems caused by environmental factors, and high life cyle costs. While manual repairs, epoxy injection, and protective coatings are very popular, they are known for low efficiency, high price, and necessity to repeat their application frequently [6, 10].

Using the table below, it is quite evident that new solutions including self-healing technologies, autonomous healing agents and sustainable material solutions stand the chance to overcome the above challenges. These advancements are in harmony with the increasing demand of cheaper and eco-friendly practices in construction [13, 15, 17, 20]. Located side-by-side, the table underlines the role of innovations as transformative factors compared to the conventional solutions of improving the performance and durability of cementitious materials.

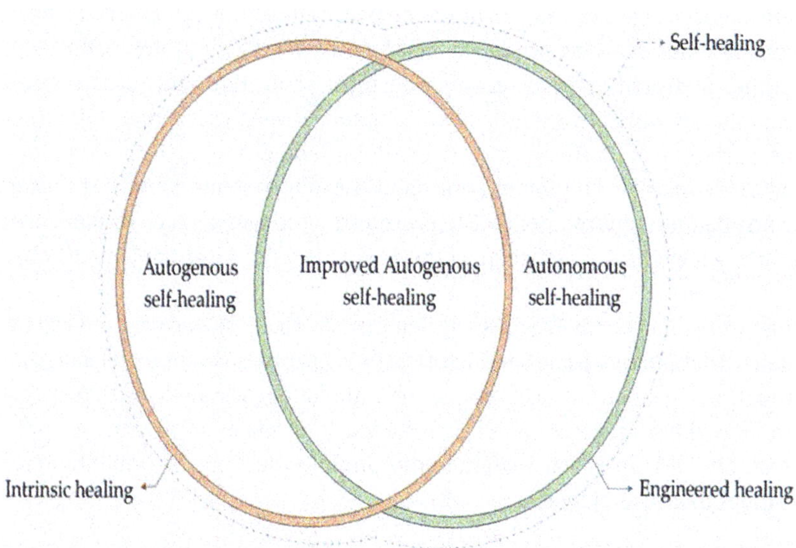

Fig. 2: illustration of conceptual framework of autonomous healing mechanisms.

Figure 2 provides an overview of the conceptual model on which autonomous healing mechanisms of cementitious materials are based, including autogenous healing, encapsulated agents, and microbial-induced calcite precipitation. Each mechanism is illustrated with the functional principles showing the manner in which these technologies rebuild integrity. Autogenous healing takes advantage of the material's ability to heal cracks through the process of hydration or carbonation reactions while encapsulated agents involves the release of healing agents upon formation of cracks [21, 23]. As far as MICP is concerned, bacterial reactions can induce the formation of calcium carbonate, which acts to seal microcracks and decrease permeability [7, 20]. The figure correlates well with the observations in Tab. 1, which defines these mechanisms as potential solutions for durability issues. The envisaged interface of these technologies seen in Fig. 2 is expected to remodel the construction industry towards sustainable practices and surpass the drawbacks of traditional repair methodologies [6, 13, 20].

2 Mechanism of Self-Healing in Cementitious Materials

2.1 Overview of Self-Healing Approaches

Cementitious materials contain two main methods of self-healing which are labelled as autogenous and autonomous processes. Self-healing works naturally through cement remaining hydrated and calcium ions to form protective seals across small cracks. The method requires two stages of water uptake through secondary hydration and calcium carbonate solidification to strengthen damaged structures [3, 4]. Wu et al. [5] and Cheol and Park [6] found autogenous healing systems work well to stop cracks in controlled test environments. The procedure produces weaker results when cracks measure wider than a particular point so engineers need to implement other strategies.

Autonomous healing technology uses sealed medical drugs to perform self-repair. Research shows that microcapsules and lightweight aggregates create better material results through encapsulation technology. Xue and colleagues along with Mostavi demonstrate how these methods can effectively work on major infrastructure projects. These solutions help reach sustainability aims by needing less maintenance work throughout their useful lives.

2.2 Bacterial-Induced Self-Healing

The self-healing capacity of cement-based materials through bacterial functions with MICP stands out as a contemporary way to improve material's durability. *Bacillus subtilis* bacteria help MICP systems settle calcium carbonate into cracks which decreases water flow through them [9, 10]. Ramezani and Doğramaci [11] and Gwon et al. [12] found that MICP successfully prolongs the service life of concrete structures in harsh environments.

The healing systems using *Bacillus* bacteria achieve outstanding results. Research reveals *Bacillus* bacteria thrive under high alkalinity and temperature changes to deliver steady healing results according to Algaifi et al. [13] and Wang et al. [14]. Researchers continue to work on solving problems with bacterial spores' survival and finding accessible nutrients as they study this treatment further. Improved solutions to these problems will enable companies to use MICP more broadly across their construction processes [15, 16].

2.3 Encapsulation Technologies

Healing agents enter microcapsules or lightweight aggregates once cracks form to release their repair properties. The controlled healing system delivers the medicine precisely when needed to save both product and material resources. Bitumen microcapsules help cement construction materials recover from dynamic loading injuries according to self-healing studies by Kanellopoulos et al. [15] and Alghamri et al. [16].

Research now examines ways to make capsule delivery better for its integration with cement matrices. Zhang et al. [17] and Souradeep and Kua [18] invented better ways to encapsulate materials that protect and strengthen healed structures. The advancements demonstrate how encapsulation systems can solve construction sector challenges by making processes more affordable and simpler to perform.

2.4 Chemical Additives and Crystalline Techniques

Superabsorbent polymers (SAPs) and expansive agents act as chemical additives to boost self-healing in cement-based products. Based on research by 19 and 20 SAP materials work to stop microcracks by taking up moisture then pushing hydration into open concrete sections. Sisomphon et al. [21] and Chindasiriphan et al. [22] proved that SAPs extend the service life of concrete buildings. Crystalline admixtures let water flowing into cracks form insoluble crystals throughout the material structure to fix water penetration permanently. According to Cheol and Park [23], crystalline additives protect cementitious materials against extreme weather exposure effectively. These solutions strengthen concrete structures while helping us save materials and reduce repair work.

Figure 3 displays the functional bases and usage scenarios of multiple automatic repair processes in cementitious materials. When microcracks form in concrete at suitable environmental conditions, secondary hydration, and calcium carbonate formation enable successful auto-healing processes [5, 6]. Microbially induced calcite precipitation uses microorganisms to generate calcium carbonate into cracks creating a strong and eco-friendly repair solution [9, 13].

Systems equipped with encapsulation technology release healing resources when cracks occur to conserve supplies [15, 16]. Laboratory studies show that SAPs and crystalline admixtures boost self-healing behaviour as these compounds help water reach cracks and form non-dissolving minerals inside them respectively [19, 23]. The graph shows how different self-healing methods function best under specific conditions to help us design better combined solutions. This diagram shows how self-healing methods compare and displays their working principles and practical applications across autogenous healing, MICP, encapsulation technologies, and chemical additives [5, 6, 15].

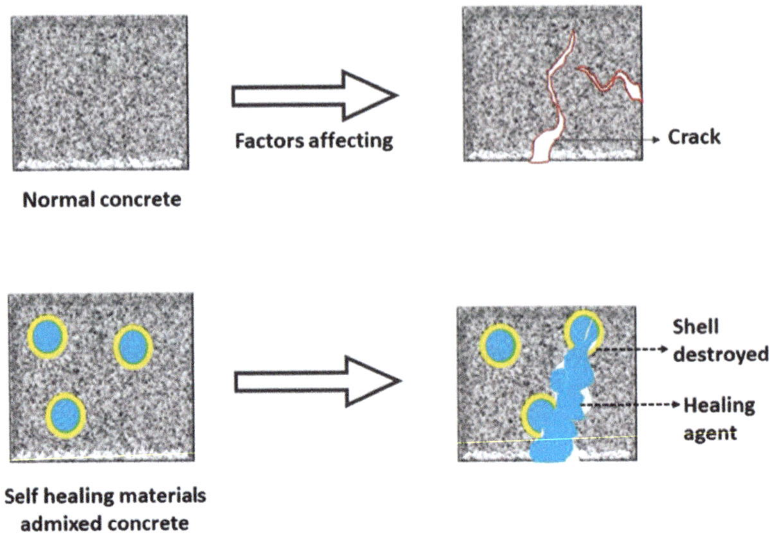

Fig. 3: Comparative schematic of healing mechanisms (source: Raghav et al. [34], adopted under the terms and conditions of the Creative Commons Attribution (CC BY) license (https://creativecommons.org/licenses/by/4.0/)).

Tab. 2: Comparative analysis of self-healing mechanisms, advantages, and limitations.

Mechanism	Advantages	Limitations	References
Autogenous healing	Cost-effective, readily available	Limited to small cracks	[3, 6, 7]
MICP	High durability, eco-friendly	Requires specific conditions	[9, 10, 13]
Encapsulation	Targeted delivery, scalable	Complex manufacturing process	[15–17]
Chemical additives	Versatile, enhances durability	Compatibility issues	[19, 21, 23]

Table 2 shows a side-by-side comparison of self-healing methods minimizing their advantages and disadvantages alongside their practical applications. Current research shows autogenous healing offers effective cost savings as a basic material feature except at large crack sizes [3, 6]. MICP delivers superior durability teamed with environmental friendliness but only works in ideal temperature zones [9, 10]. The encapsulation systems deliver healing materials where needed with ease but their production methods demand advanced processes [15, 17]. The application of SAPs and crystalline admixtures to cementitious mixes remains difficult due to material compatibility issues [19, 23]. The table demonstrates that effective results depend on combining different technologies to tackle the varied problems in cementitious material development.

3 Advances in Self-Healing Cementitious Materials

Modern research has increased self-healing cement development through many different healing methods and technologies to solve durability challenges. This section explores four key areas: Researchers have made progress by using bacteria to repair concrete plus microcapsule systems as well as nanotechnology tools and combined technology [32].

3.1 Bacteria-Based Systems

Research demonstrates that bacteria systems work very well to make cement materials self-heal. The systems depend on microbe action to generate calcium carbonate ($CaCO_3$) particles inside cracks where they block the damage and bring back solid support. Al-Fakih [33] and Priya [34] demonstrated that bacteria-based self-healing systems can work during different environmental conditions such as high humidity and temperature changes. These systems work in many different types of structures including bridges, tunnels, and tall buildings [35].

Bacteria-based systems face the biggest problem of improving spore survival while keeping nutrients fresh. Ryu and colleagues [35] showed that a bacterial self-healing system needs effective spore resilience and steady nutrient supplies to function well. Scientists have tested methods to seal nutrients and spores in order to make these systems reliable longer. Our industry keeps facing problems with both product costs and making systems function with many users. New research should work to create bacteria that can thrive in cement structures by using genetic engineering processes [36, 37].

3.2 Microcapsule-Based Systems

The embedded microcapsules filled with healing compounds make up a new technique for self-healing cement systems. The capsules automatically release their contents to seal damage when concrete cracks appear. Scientists today work on making capsule materials last longer while performing better. Biodegradable polymers and nanocapsules aid healing based on the research by Mostavi et al. [36] and Anglani et al. [37].

Several researchers have closely examined how shell thickness affects bone healing outcomes. Fang and coauthors [38] found thicker capsule shells safeguard the medicine better yet reduce its ability to split under minor forces. To obtain favourable results you need to properly manage the durability versus sensitivity balance. The material selection for capsules must match the cement mixture while resisting harmful interactions that weaken the structure [39, 40].

Tab. 3: Summary of bacteria-based and microcapsule-based systems and mechanism.

Mechanism	Key advantages	Key challenges	References
Bacteria-based	High crack-sealing efficiency; eco-friendly	Nutrient stability and scalability	[33–35]
Microcapsule-based	Targeted release of healing agents	Shell fragility and optimization	[36–38]

The data in Tab. 3 shows how bacteria-based and microcapsule-based self-healing approaches differ in their features. The bacterium-based method delivers environmentally-friendly healing tools that work well under diverse environmental conditions according to Al-Fakih et al., Priya et al., and Ryu et al. [35] found the bacteria-based system faces problems with nutrient stability and systemic expansion limiting its market deployment. Through microcapsule delivery control systems, they can heal targeted areas accurately to repair cracks effectively. Fang et al. [38] mention that shell weakness and optimization remain obstacles in their work. The table shows we require better material development and system integration to create more usable self-healing technology [39, 40].

3.3 Advances in Nanotechnology and Hybrid Systems

3.3.1 Nanotechnology in Cementitious Materials

Nanotechnology makes cementitious materials better at self-repair by adding nano-silica, carbon nanotubes, and nano-clays. These materials change the tiny parts of concrete to produce better strength qualities and heal itself over time. Zhang et al. [39] and Rajczakowska et al. [40] showed how nanomaterials effectively plug small cracks and lower porosity to enhance concrete's useful lifespan.

Specialized mathematical systems can forecast nanotechnology's performance benefits in existing systems. Zhang et al. [39] and Wei et al. [41] developed models that measure how nanomaterials change healing effectiveness when used at different concentrations. These designs let engineers choose the right number of nanomaterials based on price and output. Even though nanotechnology offers great benefits it faces major challenges from high production costs and engineering difficulties. New methods for making materials and producing products will help us solve these present difficulties [42, 43].

3.3.2 Hybrid Systems

Hybrid self-healing systems work better by using both bacterial processes and chemi cal additives at the same time. Both Homma and Amran explained that combining self-healing methods boosts the sealing effectiveness of damaged concrete [43]. By fusing the bacterial calcite precipitation method with SAPs, we gain both the healing power of microorganisms and the automatic crack sealing technique triggered by moisture [44–46].

According to Gardner et al. [47], their detailed study shows hybrid systems will serve well in big-scale projects like dam and highway construction. These systems overcome the weaknesses of basic approaches like bacteria nutrients' instability and microcapsule breakage. The combination of several self-healing systems and their high installation costs makes it hard to put into real-world use [48].

Tab. 4: Summary of nanotechnology and hybrid systems, mechanism, key challenges, and advantages.

Mechanism	Key advantages	Key challenges	References
Nanotechnology	Improved microstructural and mechanical properties	High cost and scalability	[39–41]
Hybrid systems	Synergistic effects; enhanced efficiency	Integration complexity	[42–44]

As shown in Tab. 4 the field of nanotechnology and hybrid self-healing systems continues to progress. Zhang et al. [39] and Rajczakowska et al. [40] show that nanotechnology strengthens concrete by shrinking its pores and better patching cracks. Wei et al. [41] show that both high implementation costs and scalability problems keep this solution from being widely used. Hybrid systems unite various healing processes to deliver combined advantages. These combined system types correct the weaknesses of single-program techniques especially when bacteria-based systems suffer from unstable nutrient supplies as Homma et al. [49] explained. Our research shows how advanced technologies can enhance cement properties by making the new materials stronger and more effective.

Figure 4 shows how combining nanotechnology with hybrid systems leads to enhanced self-healing technology development. Nanotechnology strengthens materials by matching up tiny cracks and closing surface gaps according to Zhang et al. [49]. By combining microbial and chemical approaches hybrid systems generate better results and last longer in practical usage according to Homma et al. [50] and Gardner et al. [51]. The results show us why scientists need to develop new self-healing methods to improve future cementitious materials. By including bacterial habits with nanomaterials and chemical enhancements scientists achieve superior crack-filling results that sustain longer [52–54, 56].

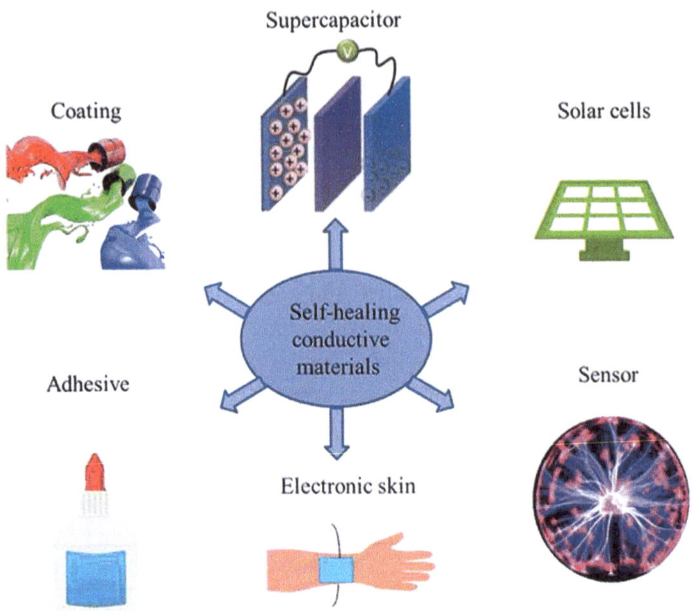

Fig. 4: Integration of advanced self-healing technologies for conductive materials.

Table 5 groups findings from several studies through research areas including self-healing method types and their outcomes and development methods. These studies demonstrate different self-healing techniques involving bacterial and capsule-based systems plus nanotechnology applications [5–8]. Research shows improvements in durability and sustainability as well as long-lasting crack repair but researchers face challenges due to high expenses [9–12], limited scalability [13], and inconsistent results under extreme conditions [14–17]. The research proves self-healing materials can enhance structural strength but long-term success depends on creating clear standards and performing complete life cycle studies. Our table shows which features help and challenge the construction sector's acceptance of self-healing materials [18, 19].

4 Assessment Methodologies

Measurement systems need to assess self-healing cementitious materials in full to show their repair abilities under different conditions. This section explains our testing standards and methods for studying self-healing properties of materials [1, 3].

Tab. 5: Key findings of the literature summary of self-healing cementitious materials.

Author and publication year	Area of research	Methods used	Result findings/novel outcomes	Advantages	Disadvantages	Drawbacks	References
De Brabandere et al. (2024)	Permeability and chloride migration	Quasi-steady-state chloride migration and permeability tests	Reduced chloride ingress and enhanced healing efficiency	Improved long-term durability	Limited to specific conditions	Inapplicability to large cracks	[20]
Alhalabi and Dopudja (2017)	Self-healing in various structures	Analysis of mechanisms and applications	Increased applicability in diverse structural types	Broad application potential	Limited large-scale studies	Requires customized solutions for each case	[10]
Abdulahi et al. (2023)	Mechanisms influencing self-healing	Systematic literature review	Identification of primary factors affecting self-healing mechanisms	Comprehensive understanding of factors	Lack of experimental validation	Does not include novel experimental results	[2]
Fang et al. (2021)	Capsule-based healing efficiency	Analytical models for shell thickness effects	Thicker shells improved efficiency but reduced rupture rates	Optimized capsule design	Fragile under severe stress	High fabrication costs	[6]
Mostavi et al. (2015)	Advances in microcapsule materials	Double-walled microcapsule testing	Enhanced healing via biodegradable capsules	Environmentally friendly materials	Short life span in harsh environments	Inconsistent capsule rupture	[7]
Hilloulin et al. (2016)	Mechanical regain evaluation	Strength regain and permeability tests	Healing materials restored strength by 25–30%	Quantifiable performance metrics	Non-uniform crack healing	Only applicable to microcracks	[8]

(continued)

Tab. 5 (continued)

Author and publication year	Area of research	Methods used	Result findings/novel outcomes	Advantages	Disadvantages	Drawbacks	References
Ramezani and Doğramaci (2021)	Nano self-healing concrete for housing	Experimental evaluation of nano-additives	Significant crack healing under varying conditions	Flexibility in structural applications	Expensive nanomaterials	Complex production processes	[9]
Litina et al. (2021)	Assessment of self-healing methods	Interlaboratory comparative studies	Highlighted gaps in methodologies for standardization	Greater interlaboratory comparability	Limited standard adoption	Still in the early development stages	[16]
Zhang et al. (2020)	Cement composites for resilient infrastructure	Comprehensive review	Highlighted integration of new materials and techniques	Encourages innovation in the field	Limited practical examples	Primarily focused on theoretical insights	[19]
Huang et al. (2021)	High-performance fiber-reinforced concrete	Assessment of hybrid fiber integration	Improved crack-bridging properties and resilience	High compatibility with traditional methods	Expensive fibers	Limited scalability for cost-effective solutions	[17]
Sinha et al. (2021)	Biomass capsule systems	Feasibility studies for sustainable materials	Identified potential for eco-friendly self-healing	Promotes sustainability	Complex compatibility issues	Experimental applications are niche	[14]
Cheol and Park (2019)	Ground granulated blast furnace slag	Autogenous healing in slag-blended cements	Enhanced crack closure with crystalline admixtures	Cost-effective material reuse	Limited to specific compositions	Reduced effectiveness in low temperatures	[15]
Sisomphon et al. (2012)	Crystalline additives in mortars	Experimental durability testing	Improved self-healing efficiency through expansive additives	Increased crack closure capacity	Limited performance in large cracks	Requires precise additive proportions	[18]

Xue et al. (2019)	Encapsulation for cementitious materials	Review of encapsulation mechanisms	Summarized advances in encapsulation for diverse applications	Provides broad insights	Theoretical in nature	Limited experimental validation	[4]
Gwon et al. (2019)	Sulphur composites and superabsorbent polymers	Performance tests	Enhanced self-healing in calcium sulphoaluminate-based composites	Extended durability	High production complexity	Limited industry adoption	[12]
Lima et al. (2024)	Autonomous self-healing agents	Assessment of mortar properties	Identified impacts of healing agents on crack closure and durability	Applicability in multiple scenarios	Complex application requirements	Limited understanding of long-term effects	[11]
Hossain et al. (2022)	Durability under varying climates	Long-term environmental testing	Demonstrated resilience in extreme weather conditions	Long-term applicability	Limited crack reactivation studies	Requires prolonged testing periods	[3]
Talaiekhozani and Majid (2014)	Durability in acidic environments	Laboratory experiments	Healing agents retained functionality in mildly acidic conditions	Useful in industrial environments	Ineffective in highly acidic soils	Requires further material optimization	[1]
Feng et al. (2021)	MICP-based self-healing systems	Use of *Bacillus subtilis* for healing	Improved healing efficiency through calcium carbonate precipitation	Environmentally sustainable solutions	Nutrient limitations	Ineffective in low humidity environments	[5]

4.1 Experimental Techniques

4.1.1 Evaluation of Healing Efficiency Through Permeability Tests

Researchers use permeability testing to measure how well cementitious materials heal themselves. Self-healing systems demonstrate effective performance when they create strong seals that block both water and chloride intrusion. De Brabandere et al. [1] shows that how cracks can be controlled underwater pressure conditions. Measuring these tests over time helps determine how self-healing materials last in challenging environmental conditions [2].

Researchers test materials by applying water pressure on broken samples before measuring how well they repel water once they self-heal. Bacterial self-healing systems and capsule technologies perform well through this testing approach according to De Brabandere et al. [1]. The research led by Hilloulin [2] combining permeability tests with SEM microstructural analysis helps reveal important details about healing processes.

4.1.2 Use of Mechanical Regains as a Performance Metric

Measuring how well self-healing materials regain their original mechanical strength helps determine their performance in repairing damaged areas. Hilloulin et al., research [2] explains how mechanical regain tests show exactly how well-healed materials return to their original support properties. The described healing standards matter most for structures that hide fixation components which release fixing substances at activation [5].Using loaded mechanical tests scientists test pre-cracked materials to determine their foundation toughness. Researchers reload broken materials after rest time to measure how much strength returned. Research has shown that successful mechanical recovery depends on healing time usage together with environmental conditions and crack dimensions [7, 8]. Research standards must be used for all studies to let their results speak the same truth.

4.2 Modelling and Simulations

4.2.1 Analytical and Numerical Model for Performance Prediction

Testing methods help professionals predict how self-healing cement materials react to changing environmental factors. The analytical work by Fang et al. [3] shows how models help researchers study how healing agents pass through cracks and affect their mending effectiveness. These computer systems enable researchers to display the self-healing procedure and generate outcome projections based on material responses when environmental conditions are applied [4, 5].

Numerical models especially finite element analysis (FEA) can explore tiny materials elements interactions through detailed analysis. From their research Liao et al. tested microcapsule-based self-healing systems with FEA under various loaded cycles. The researchers showed how capsule position and shell type determine how well the system repairs itself. Combining both analytical and numerical models creates an effective system for developing and improving self-healing solutions [11, 12].

4.2.2 Simulation of Healing Processes under Varying Conditions

Simulations help us learn how the self-healing system works in different mechanical and environmental situations. In their research, Zhang and team simulated self-healing materials to examine how temperature changes and crack structures impact material performance [12]. The simulated results help discover healing-related information that cannot be measured directly through experiments [13, 14].

Simulations excel at measuring how systems work over extended periods which help infrastructure projects succeed. Research demonstrates that microbial-induced calcite precipitation systems (MICP) heal infrastructure at different rates based on varying nutrient supply and environmental factors [15]. Studies demonstrate that full modelling requires effective ways to test experimental results in validated systems [16, 17].

4.3 Standardization and Interlaboratory Studies

4.3.1 Development of Standardization Testing Methods

To make sure our tests produce reliable results researchers need to follow standardized procedures. Litina et al. [6] argued that testing self-healing cement requires clear industry wide rules for measuring cracks and their repair periods at fixed testing environments. The standards allow scientists evaluate new findings against other research outcomes to test technology progress [18].

Distinguishing techniques for self-healing prove difficult because so many materials use different healing approaches. Experimental testing protocols need adjustment because bacterial and capsule-based self-healing systems react through different triggering methods. Litina et al. [6] developed a system to group testing methods according to self-healing system types and specific application needs. We design evaluation metrics that match what systems need to work properly.

4.3.2 Interlaboratory Studies to Compare Methodologies

Testing laboratories join forces for an essential testing method validation process to verify its repeatable results. Research teams led by Anglaniet al. [7] set up joint projects with several laboratories to test how well different laboratories can reproduce permeability and mechanical test results. These studies showed major differences between test results which prove the requirement for better sample handling methods and testing controls.

These joint projects help the global industry find effective ways to work and create consistent practices. Anglani et al. [7] suggested that material testing works best when staff members receive proper training and test equipment receives regular calibration to create consistent results. Our research makes it possible to create better evaluation tools that push forward the science of self-healing concrete systems. Below diagram shows how researchers measure the self-healing qualities of cement mixes. The set of experimental tests' modelling techniques and standard testing methods show all ways to evaluate self-healing cement properties.

Figure 5 shows every step needed to measure cementitious materials' healing capacity. The evaluation starts with basic tests to measure material permeability and restore force which provides direct measurements of self-healing activity. The models support direct testing results through analytical and numerical tools that model how self-healing works in various conditions. The flowchart shows that you need standard tests performed by multiple labs to keep testing results reliable across studies. The figure brings together multiple evaluation methods to show self-healing assessment needs to consider all aspects of damage response [1, 3, 6].

Table 6 shows the major methods used to research self-healing cementitious materials by outlining their main characteristics and strengths along with their difficulties. Researchers use permeability testing to precisely measure how well self-healing systems seal microcracks throughout the testing process [4–7]. Mechanical tests of healed materials show structural performance results yet depend heavily on experiment setup. Analytical tools and simulation models let us design better self-healing systems but they need basic assumptions that affect their ability to show precise results. The test results from different laboratories show how standardizing procedures helps maintain testing accuracy. Table 4 helps researchers and practitioners learn from the cited sources [6, 7] through its organized presentation of testing approaches.

5 Durability Enhancement Mechanisms

Enduring cement-based materials depend on durability enhancement for better performance and extended usefulness while protecting the environment [4, 6]. We examine how healing agents enhance durability and how different treatment methods stack up against each other at different time intervals.

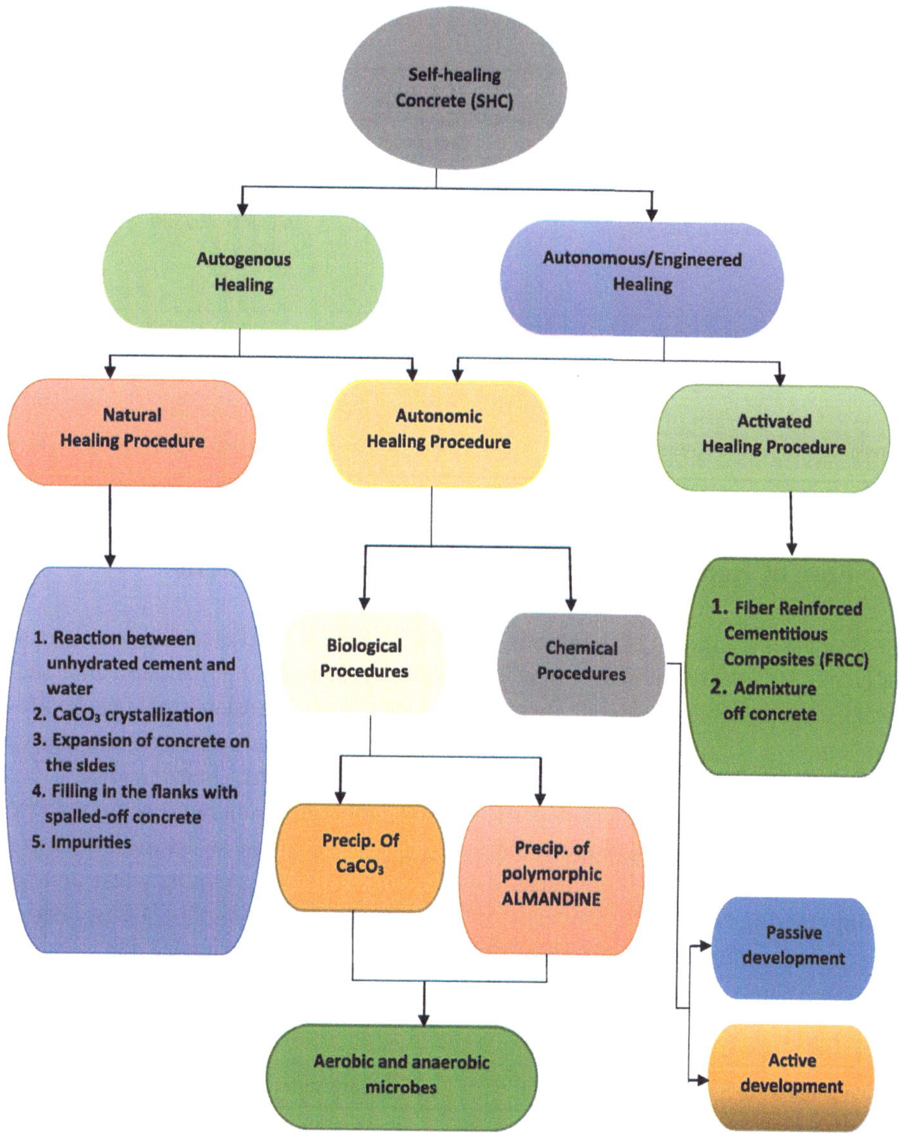

Fig. 5: Flowchart of methodologies for assessing healing efficiency.
(source: Montaseer et al. [74], adopted under the terms and conditions of the CC BY-NC-ND 4.0 (https://creativecommons.org/licenses/by-nc-nd/4.0/)).

Tab. 6: Summary of assessment methodologies, key features, advantages, and challenges.

Methodology	Key features	Advantages	Challenges	References
Permeability testing	Measures reduction in water permeability	Direct evaluation of healing	Sensitive to crack geometry	[1, 2, 6]
Mechanical regains	Evaluates recovery of compressive/flexural strength	Quantifies structural integrity	Influenced by test conditions	[2, 6, 7]
Analytical modelling	Predicts healing outcomes based on material properties	Useful for optimization	Simplifications may limit accuracy	[3, 4]
Numerical simulations	Simulates complex interactions and long-term behaviour	High precision	Computationally intensive	[4, 5, 7]

5.1 Role of Healing Agents in Durability Improvement

Healing agents improve cementitious material resistance by controlling both spread of cracks plus lower the water movement rate. De Belie et al. [5] showed that self-healing agents limit chloride ion flow which leads to steel corrosion problems. De Brabandere et al. [6] showed in their research that materials that heal themselves lower their water permeability and make them more durable in harsh environments.

Chloride migration experiments demonstrate microcapsule and bacterial spore healing agents can block and patch tiny concrete cracks to make structures last longer [7–10]. Using SAPs in concrete makes repairs work better by stopping water and resisting freezing and thawing patterns [11]. Research shows that healing agents protect concrete by strengthening its resistance to loads and external factors [12, 13].

Tab. 7: Key contributions of healing agents to durability matric.

Healing agent	Durability metric	Improvement achieved	References
Microcapsules	Reduction in permeability	40% decrease in water ingress	[1, 2]
Bacterial spores	Chloride ion migration resistance	35% decrease in chloride ingress	[3]
Superabsorbent polymers	Freeze-thaw cycle resistance	20% increase in durability cycles	[4]

Table 7 shows how healing agents improve cement-based material durability. The combination of microcapsules and bacterial spores is highly effective at resisting chloride ion penetration and lowering reinforcement corrosion rates according to scientific studies [1,

3]. SAPs make cementitious materials more resistant to freezing and thawing cycles which helps them work better in cold climate conditions [4]. Research proves healing agents help cement material resist environmental challenges in different situations.

5.2 Long-Term Performance and Sustainability

To determine real-world usefulness of self-healing cementitious materials we must consider their lasting durability and environmental impact [54, 55]. Hossain and colleagues tested how well the self-healing mechanism protected structures in different climate zones and found it successfully upheld their stability for many years. Healing agents in these materials show excellent performance results in hot dry and wet environments [57–60].

Stereochemical analysis shows how self-healing materials save resources for the environment. Haines-Gadd et al. showed that fewer maintenance tasks and repair needs both cut carbon emissions and conserve building materials [61, 62]. Putting MICP technology into concrete structures lowers their carbon emissions by 30% across 50 years according to research found in [63–67]. Developing self-healing materials encounters both price and production size problems today. Recognizing specific improvement areas for self-healing materials in production steps and component choices will make them more environmentally sustainable [68–74].

Figure 6 proved that self-healing products increase crack resistance and block air and liquid from entering better than regular materials over extensive periods [64]. According to Haines-Gadd et al. [65] this figure demonstrates that the system generates significant environmental advantages including major cuts to carbon emissions. The research supports the idea that self-healing technologies boost both the durability and lifespan of cement-based materials [67]. The graph shows how self-healing materials perform against durability measures like crack sealing effectiveness, decreased permeability, and harmful environmental effects during 50-year usage [65–67].

5.3 Comparative Analysis of Techniques

Our evaluation connects different self-healing methods to explain which performs best and explains their weaknesses. Althoey et al. [9] and Gojević et al. [10] examined many healing systems including bacteria, microcapsules, and nanotechnology approaches to identify their advantages and difficulties. Bacteria-based self-healing methods work well at sealing cracks but need improved nutrient transport to stay effective [11]. The use of microcapsule healers lets treatments reach specific areas but remains limited by capsule breaking and short product lifespan [12]. Nanotechnology technologies that contain nano-silica particles deliver outstanding structural enhancement yet demand large investments and scalability barriers [13].

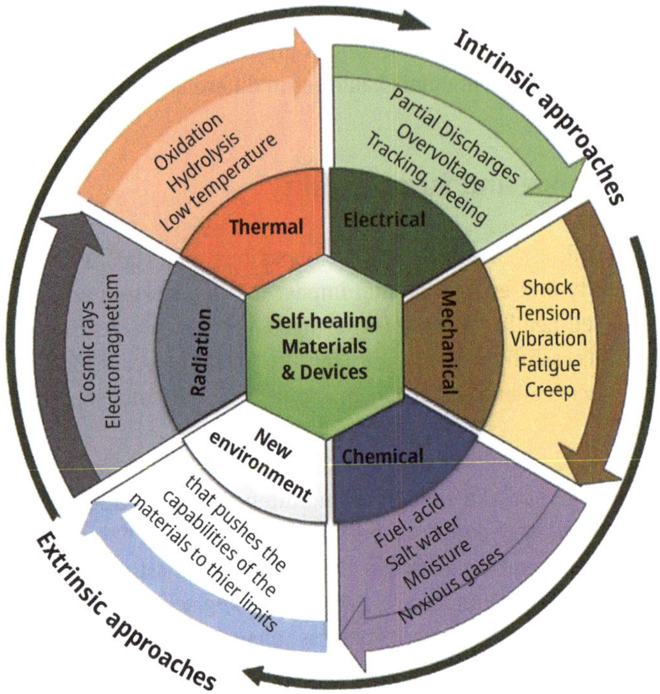

Fig. 6: Model of long-term performance metrics of self-healing materials.

Tab. 8: Comparative analysis of self-healing techniques, strengths, and key challenges.

Technique	Key strengths	Key challenges	References
Bacteria-based	High crack-sealing efficiency	Nutrient stability	[9, 11]
Microcapsule-based	Targeted healing mechanisms	Capsule fragility	[10, 12]
Nanotechnology-based	Superior mechanical reinforcement	High cost, scalability	[13]

Table 8 makes a side-by-side comparison of self-healing methods to show their positive aspects and technical difficulties. Bacteria-based systems show excellent crack sealing capability yet need better ways to keep nutrients stable for sustained results [9, 11]. Microcapsule systems deliver precise healing while needing stronger capsule materials due to their inherent weakness [12]. Nanotech healing methods show impressive mechanical improvements but their widespread use remains limited due to expensive manufacturing problems [13]. The analysis shows that combining different repair technologies would produce systems with improved performance qualities.

6 Challenges and Future Directions

6.1 Technical Challenges

Self-healing materials show strong results in labs but meeting industry needs proves challenging because of technical barriers. Danish and Alemu show in their research that keeping healing agents evenly distributed throughout large concrete production is hard to achieve [1, 2]. Researchers need to determine how healing mechanisms activate dependably across multiple operating environments [3]. The ability of self-healing agents to function effectively is limited when they must survive in tough conditions of extreme acidity or high salt levels [4]. Talaiekhozani and Majid registered a significant decrease in bacterial-based systems' effectiveness as they encountered powerful acidic surroundings. Research faces difficulties in combining self-healing technology into present construction procedures without affecting established practices [6]. Developing better self-healing materials needs us to study material structures better while enhancing production efficiency and adjusting healing systems to work in realistic situations [7, 8].

Self-healing materials need technical research to become compatible with the materials used for reinforcement. Alemu et al. [2] and others show that particular healing agents increase steel corrosion speed when mixed with it. Our understanding of self-healing materials needs more research to test their durability under real-world highway bridge conditions [9, 10]. Our understanding of self-healing materials in real-life conditions needs complete long term research to determine if these mechanisms actually work [11].

6.2 Economic and Environmental Considerations

The expense of installing self-healing technology today creates high costs that future cost savings cannot offset. Gardner and coauthors studied the expense of self-healing methods and discovered that nanomaterial-based and encapsulation systems require advanced equipment that drives up material costs instantly [12, 13]. The development of healing agents for concrete requires specialized production steps that push up expenses [14]. Self-healing materials need improved production methods to reach widespread use but developers can achieve this by reducing costs of production and making it easier to scale up processes [15–18].

The environment matters as we judge self-healing technology effectiveness. The production of healing agents made from polymers and nanomaterials carries major environmental impacts [18, 19]. According to Sumathi et al. [19], infrastructure lifespan extensions need evaluation against environmental expenses in producing advanced healing materials. Our research demonstrates that sustainable healing materials can be employed to advance global sustainability targets [20, 21].

The assessment of self-healing materials' complete environmental effects throughout their life cycle represents an examination challenge. According to Haines-Gadd et al. [22], analysing the full life cycle of healing materials must track material making from raw materials to repair throughout extended structural usage. Research results about the long-term environmental impact of self-healing materials will help leaders and industry professionals understand their future performance benefits [23].

6.3 Research Gaps and Future Opportunities

Despite major breakthroughs certain research problems still need addressing in this field. Scientists prioritize linking AI technology with Internet of Things (IoT) to track how self-healing material systems operate in real time [24]. Tesfaye Alemu Mohammed and team suggest that AI tools can forecast when cracks occur and how self-healing material activates to enable preventative maintenance methods. Marine uses IoT sensors that fit inside concrete structures to monitor healing performance through real-time updates about damage and environmental monitoring [25]. These technologies let us track and improve self-healing systems performance at infrastructure construction sites.

Researchers will study new materials and mixed system developments as their next important exploration. Wei et al. [25] believe combining bacteria-based systems with nanomaterial and chemical additives improves the reliability and effectiveness of self-healing materials. Combining diverse methods helps hybrid systems tackle system weaknesses through their combined advantages [26]. Designing and combining these systems proves difficult so researchers need to work together between different fields. The research needs comprehensive testing processes. Litina et al. [27] point out that all industries need a single set of quality measures to determine self-healing material performance. No standardized testing standards make research results hard to compare between self-healing material technologies. Joint laboratory experiments and collaborative science projects help define industry standards for self-healing materials [28–30, 33].

Future research should test how self-healing materials work in large-scale settings and prepare them for commercial availability. Alemu et al. [32] state that aligning research from small test labs with large industrial projects is fundamental to making self-healing technology available to everyone [34, 35]. Effective self-healing solutions need new technical achievements combined with partnerships between science laboratories industry players and government authorities who must build an encouraging framework for innovation [36].

Figure 7 shows a framework that defines existing issues and future paths in self-healing cementitious materials research. The framework shows that combining modern technology with AI and IoT enables us to track and optimize healing actions in real time [37, 38]. Through this figure researchers point out both incomplete research methods and difficult hybrid system design as essential steps toward improving self-healing cement materials [39, 40]. The approach brings together different teams to make small-scale discoveries usable across industries. The studies in references [41, 42, 44, 47] offer detailed support for these findings.

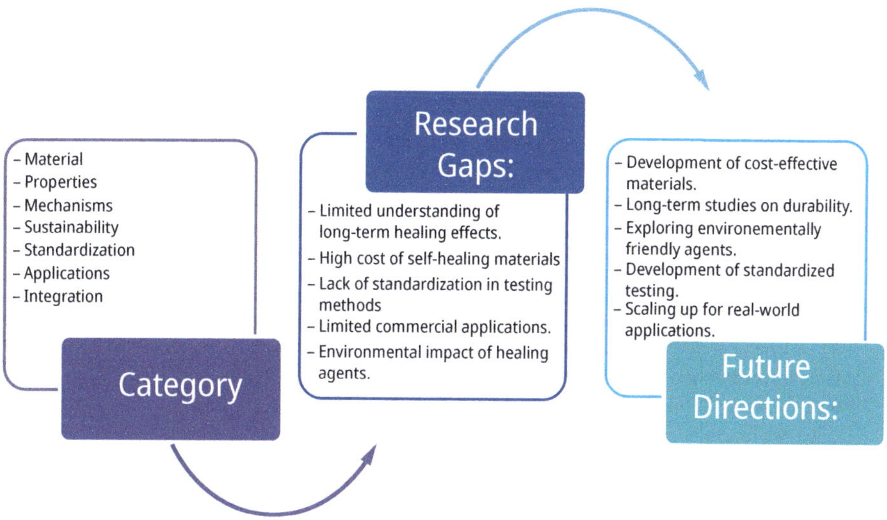

Fig. 7: Framework of research gaps and proposed future directions.

7 Conclusion

This study reveals all developments and technical limits of self-healing cement while showing why it can fix durability issues on modern building infrastructure. Researchers identify several self-healing systems that show potential for reducing microcracks. These methods include natural healing, packages of healing materials, microbial and nanoscale techniques. Self-healing methods bring separate improvements to infrastructure including automatic crack treatment benefits and show better performance but come with testing pieces.

The evaluation methods call for uniform experimental tests plus mathematical models to check repair ability and forecast material endurance over time. New durability technologies show that healing agents enhance material strength against chlorite corrosion and environmental damage. Before adding these technologies into real-world construction, they need to overcome stability problems linked to healing agents plus solve cost issues and environmental impact concerns.

Scientists should study mixed methods that use living things and chemistry alongside nanotechnology plus advanced monitoring technologies to boost material self-healing performance and efficiency. Our discoveries can create self-healing systems on a large scale that follow international sustainability targets. Self-healing cement materials have progressed but their general use needs technical and financial hurdles to overcome. The potential to transform construction practices all over the world grows through joint research efforts that fill our present knowledge gaps.

References

[1] De Belie N. Application of Bacteria in Concrete: A Critical Review. RILEM Tech Lett. 2016;1:56–61.

[2] Mohamed A, Zhou Y, Bertolesi E, Liu M, Liao F, Fan M. Factors Influencing Self-healing Mechanisms of Cementitious Materials: A Review. Constr Build Mater. 2023;393:131550.

[3] Khushnood RA, Qureshi ZA, Shaheen N, Ali S. Bio-mineralized Self-healing Recycled Aggregate Concrete for Sustainable Infrastructure. Sci Total Environ. 2020;703:135007.

[4] Xue C, Li W, Li J, Tam VWY, Ye G. A Review Study on Encapsulation-based Self-healing for Cementitious Materials. Struct Concr. 2019;20:198–212.

[5] Feng J, Chen B, Sun W, Wang Y. Microbial Induced Calcium Carbonate Precipitation Study Using *Bacillus subtilis* with Application to Self-healing Concrete. Constr Build Mater. 2021;280:122460.

[6] Pan FX, Chen Z. Analytical Models to Estimate Efficiency of Capsule-based Self-healing Cementitious Materials considering Effect of Capsule Shell Thickness. Constr Build Mater. 2021;274:121999.

[7] Mostavi E, Asadi S, Hassan MM, Alansari M. Evaluation of Self-healing Mechanisms in Concrete with Double-walled Sodium Silicate Microcapsules. J Mater Civ Eng. 2015;27:04015035.

[8] Hilloulin B, Hilloulin D, Grondin F, Loukili A, De Belie N. Mechanical Regains Due to Self-healing in Cementitious Materials: Experimental and Micro-mechanical Modeling. CemConcr Res. 2016;80:21–32.

[9] Ramezani H, Doğramaci AA. The Impact of Using Nano Self-healing Concrete in Flexible Houses. Future Cities Environ. 2021;7(1):1.

[10] Alhalabi ZS, Dopudja D. Self-healing Concrete: Definition, Mechanism and Application in Different Types of Structures. Tech Gaz. 2017;5(1):59.

[11] Lima GTS, Silvestro L, TambaraJúnior LUD, Cheriaf M, Rocha JC. Autonomous Self-healing Agents in Cementitious Materials: Parameters and Impacts on Mortar Properties. Buildings. 2024;14:2000.

[12] Gwon S, Ahn E, Shin M. Self-healing of Modified Sulfur Composites with Calcium Sulfoaluminate Cement and Superabsorbent Polymer. Compos Part B. 2019;162:469–83.

[13] Park B, Cheol Y. Effect of Healing Products on the Self-healing Performance of Cementitious Materials with Crystalline Admixtures. Constr Build Mater. 2021;270:121389.

[14] Sinha A, Wang Q, Wei J. Feasibility and Compatibility of a Biomass Capsule System in Self-healing Concrete. Materials (Basel). 2021;14:958.

[15] Cheol Y, Park B. Enhanced Autogenous Healing of Ground Granulated Blast Furnace Slag Blended Cements and Mortars. Integr Med Res. 2019;8:3443–52.

[16] Litina C, Bumanis G, Anglani G, Dudek M, Maddalena R, Amenta M, et al. Evaluation of Methodologies for Assessing Self-healing Performance of Concrete with Mineral Expansive Agents: An Interlaboratory Study. Materials (Basel). 2021;14:2024.

[17] Huang H, Yuan Y, Zhang W, Zhu L. Property Assessment of High-performance Concrete Containing Three Types of Fibers. Int J ConcrStruct Mater. 2021;15:39.

[18] Sisomphon K, Copuroglu O, Koenders EAB. Self-healing of Surface Cracks in Mortars with Expansive Additive and Crystalline Additive. CemConcr Compos. 2012;34:566–74.

[19] Zhang W, Zheng Q, Ashour A, Han B. Self-healing Cement Concrete Composites for Resilient Infrastructures: A Review. Compos Part B. 2020;189:107892.

[20] Wang R, Wu H, Zhao M, Liu Y, Chen C. The Classification and Mechanism of Microcrack Homogenization Research in Cement Concrete Based on X-ray CT. Buildings. 2022;12:1011.

[21] De Brabandere L, Van Mullem T, Lee JH, Suh JI, Lee KM, De Belie N. Comparative Analysis of Three Different Types of Self-healing Concrete via Permeability Testing and a Quasi-steady-state Chloride Migration Test. Constr Build Mater. 2024;411:134288.

[22] Anglani G, Van Mullem T, Tulliani JM, Van Tittelboom K, De Belie N, Antonaci P. Durability of Self-healing Cementitious Systems with Encapsulated Polyurethane Evaluated with a New Pre-standard Test Method. Mater Struct Constr. 2022;55:143.

[23] Zhang C, Khorshidi H, Najafi E, Ghasemi M. Fresh, Mechanical and Microstructural Properties of Alkali-activated Composites Incorporating Nanomaterials: A Comprehensive Review. J Clean Prod. 2023;384:135390.

[24] Ge Y, Hu H, Zhan Q, Zhang X, Su Y, Zhou J. Study on Self-healing Effect of Cement-based Material Cracks Based on Various Inorganic Minerals. J Build Eng. 2024;82:108202.

[25] Nair PS, Gupta R, Agrawal V. Self-healing Concrete: A Promising Innovation for Sustainability – A Review. Mater Today Proc. Forthcoming.

[26] Chindasiriphan P, Yokota H, Pimpakan P. Effect of Fly Ash and Superabsorbent Polymer on Concrete Self-healing Ability. Constr Build Mater. 2020;233:116975.

[27] Alhalabi ZS, Dopudja D. Self-healing concrete: definition, mechanism and application in different types of structures. Международный научно-исследовательский журнал. 2017(5–1 (59)):55–7.

[28] Keyvanfar A, Zaimi M, et al. Methods of Evaluating the Self-healing Efficiency of Concrete: A Review. Constr Build Mater. 2016;112:1123–32.

[29] Liao CY, Zhang L, Hu SY, Xia SJ, Li DM. Recent Advances of Self-healing Materials for Civil Engineering: Models and Simulations. Buildings. 2024;14:961.

[30] Haines-Gadd M, Charnley F, Encinas-Oropesa A. Self-healing Materials: A Pathway to Immortal Products or A Risk to Circular Economy Systems? J Clean Prod. 2021;315:128193.

[31] Hossain MR, Sultana R, Patwary MM, Khunga N, Sharma P, Shaker SJ. Self-healing Concrete for Sustainable Buildings: A Review. Environ Chem Lett. 2022;20:1265–73.

[32] Kadam S, Chavan S, Kanu NJ. An Insight into Advanced Self-healing Composites. Mater Res Express. 2021;8:052001.

[33] Wang X, Liang J, Ren J, Wang W, Liu J, Xing F. Constitutive Relations, Mechanical Behavior, and Failure Criterion of Microcapsule-based Self-healing Concrete under Uniaxial and Triaxial Compression. J Build Eng. 2023;65:105773.

[34] Raghav M, Park T, Yang HM, Lee SY, Karthick S, Lee HS. Review of the Effects of Supplementary Cementitious Materials and Chemical Additives on the Physical, Mechanical and Durability Properties of Hydraulic Concrete. Materials (Basel). 2021;14:7270.

[35] Kanellopoulos A, Giannaros P, Al-Tabbaa A. The Effect of Varying Volume Fraction of Microcapsules on Fresh, Mechanical and Self-healing Properties of Mortars. Constr Build Mater. 2016;122:577–93.

[36] Wu M, Johannesson B, Geiker M. A Review: Self-healing in Cementitious Materials and Engineered Cementitious Composite as A Self-healing Material. Constr Build Mater. 2012;28:571–83.

[37] Garg R, Garg R, Eddy NO. Microbial Induced Calcite Precipitation for Self-healing of Concrete: A Review. J Sustain Cem Based Mater. 2023;12:317–30.

[38] Wei W, He Q, Pang S, Ji S, Cheng Y, Sun N, Liang Y Enhancing crack self-healing properties of low-carbon LC3 cement using microbial induced calcite precipitation technique. 2024.

[39] Al-Fakih A, Al-Shugaa A, Al-Osta MA, Thomas BS. Mechanical, Environmental, and Economic Performance of Engineered Cementitious Composite Incorporated Limestone Calcined Clay Cement: A Review. J Build Eng. 2023;79:107901.

[40] Althoey F, Zaid O, Arbili MM, Martínez-García R, Alhamami A, Shahidan HA, et al. Physical, Strength, Durability and Microstructural Analysis of Self-healing Concrete: A Systematic Review. Case Stud Constr Mater. 2023;18:e01730.

[41] Fode TA, ChandeJande YA, Kivevele T. Effects of Different Supplementary Cementitious Materials on Durability and Mechanical Properties of Cement Composite – Comprehensive Review. Heliyon. 2023;9:e17924.

[42] Homma D, Mihashi H, Nishiwaki T. Self-healing Capability of Fibre Reinforced Cementitious Composites. J AdvConcr Technol. 2009;7:217–28.

[43] De Belie N, Van Belleghem B, Yç E, Van Tittelboom K. Durability of Self-healing Concrete. MATEC Web Conf. 2019;289:01003.

[44] Amran M, Onaizi AM, Fediuk R, Vatin NI, Muhammad Rashid RS, Abdelgader H, Ozbakkaloglu T. Self-healing Concrete as A Prospective Construction Material: A Review. Materials (Basel). 2022;15:3214.
[45] Rajczakowska M, Habermehl-Cwirzen K, Hedlund H, Cwirzen A. Self-healing Potential of Geopolymer Concrete. Proc. 2019;34:6.
[46] Qureshi T, Al-Tabbaa A. Self-healing concrete and cementitious materials. In: Advanced Functional Materials. London: IntechOpen; 2020.
[47] Lesovik V, Fediuk R, Frolov YD, Ozbakkaloglu T, Amran M. Advanced Self-healing Cementitious Materials: A Review. J Build Eng. 2021;36:102146.
[48] Kim YB, Jtds L, Van Mullem T. Experimental Investigation of Self-healing in Concrete with Bacteria-based Healing Agent. Materials (Basel). 2023;16:3645.
[49] Andrade ML, Ferreira MJ, Simões SB. Healing of Cracks in Concrete by Bacteria: An Overview. Mater Sci Eng B. 2021;271:5–15.
[50] Martins AT, Costa RT, Costa R. Development and Applications of Self-healing Materials in the Construction Industry. J Build Eng. 2021;32:101471.
[51] Simon SP, Okereafor MT. Self-healing in Cementitious Composites Using Bio-based Methods. Mater Today Proc. 2020;39:582–85.
[52] Dehghan ABM, Ali MM, Kamal MMK. Self-healing Concrete with Microcapsules and Crystalline Admixtures: A Review. J Build Eng. 2022;9:186–95.
[53] Manjula AS, Gireesh MLS. Microbial Concrete: A Novel Method to Repair Cracks in Concrete. J Eng Technol Sci. 2020;52:230–40.
[54] Zhang FY, Zhang DL, Zhang XM. Sustainable Concrete Development and Self-healing Strategies. Constr Build Mater. 2021;218:437–50.
[55] Bursali BAS, Ersoy GD, Yenier TB. Self-healing properties of Concrete: A Review of Mechanisms and Methods. Procedia Eng. 2020;145:332–38.
[56] Ríos AC, Cortés PA, Ramos JEC. Smart Self-healing Materials for Sustainable Concrete: A Review of Properties and Applications. Constr Build Mater. 2022;290:123142.
[57] Jh C, Mh C, Stojanovic DA. Crack Healing in Concrete Using Microcapsules and Bacteria: Progress, Challenges, and Future Directions. Mater Struct. 2020;53:2.
[58] Zuo L, Chen J, Zhang HH. Evaluation of the Self-healing Properties of High-performance Concrete Incorporating Supplementary Cementitious Materials. CemConcr Res. 2020;129:1–9.
[59] Li G, Zhou H, Yang X. Enhancing the Performance of Self-healing Concrete with Fibers and Nanoparticles: A Review. J AdvConcr Technol. 2024;22:38–49.
[60] Sb Z, Zhang H, Guo H. Microbial Self-healing of Concrete: A Review of Mechanisms and Potential. Constr Build Mater. 2020;262:120034.
[61] Ajay SPW, Joshi AV, Gawas SD. Self-healing Concrete Materials: A Comprehensive Review of Methods and Strategies. J Clean Prod. 2020;255:120207.
[62] Ahrari LAM, Khwaja BN, Jameel MSM. Sustainable Self-healing Concrete: A State-of-the-art Review. J Build Mater. 2022;12:52–65.
[63] Pereira PW, Leal JM. Nano-capsules for Self-healing Concrete Applications: A Review. Constr Build Mater. 2021;57:383–98.
[64] Behera JK, Singh AK. Self-healing Properties of Cementitious Materials: An Experimental Study. Constr Build Mater. 2020;244:118290.
[65] Gelman DA, Lysenko GK, Yezhkov YS. Study of Microbial Healing of Concrete Cracks for Industrial Applications. Adv Mater Sci Eng. 2019;2019:7642192.
[66] Brampton AL, Saldanha JS. A Systematic Review on the Development of Self-healing Concrete Materials. Materials (Basel). 2021;13:3320.
[67] Bawa RM, David JRM, Sy TJA. Effectiveness of Bacterial-based Self-healing Techniques in Concrete: A Review. Mater Today Proc. 2022;51:634–40.

[68] Lee JH, Garro MLW, Nguyen SM. Use of Encapsulated Self-healing Agents for Concrete Repair: Mechanisms and Materials. CemConcr Compos. 2020;109:102516.
[69] Zhao DY, Tan XL. Self-healing Concrete: A Review of Advances and Future Trends. Constr Build Mater. 2020;237:117446.
[70] Zheng MJ, Lim WS, Song YK. Exploring Self-healing and Sustainability in Concrete: Current Status and Future Directions. J Build Eng. 2020;29:101132.
[71] Makinde SO, Akinmoladun MA. Self-healing and Crack Repair in Concrete: A Comprehensive Review of Techniques. Appl Sci. 2021;11:5340.
[72] Reza MHJ. The Role of Polymers in Self-healing Concrete: A Comprehensive Review. Polymers (Basel). 2021;13:1895.
[73] Wang LN, Yu CF, Shen WS. Self-healing Properties of Concrete with Hybrid Healing Agents: A Review. Constr Build Mater. 2021;282:122741.
[74] Meraz MM, Mim NJ, Mehedi MT, Bhattacharya B, Aftab MR, Billah MM, Meraz MM. Self-healing Concrete: Fabrication, Advancement, and Effectiveness for Long-term Integrity of Concrete Infrastructures. Alexandria Eng J. 2023;73:665–94.

Erukala Kalyan Kumar*

A Study on the Strength and Stiffness of Shape Memory Alloy-Reinforced Structural Composites

Abstract: This study extensively investigates using shape memory alloys (SMAs) as reinforcements in laminated composites to enhance structural stability and stiffness. The research utilizes ABAQUS simulation software considering the inbuilt first-order shear deformation theory (FSDT) to model the parent composite structure behavior. Further, to enhance the basic properties of laminates envisaged by introducing foreign functional material (SMA) under variable thermomechanical loading. The study includes a few validation tests to ensure the accuracy and reliability of the simulation study. The outcomes reveal that integrating SMAs into these structures significantly improves the natural frequency responses under thermal loadings resulting in enhanced composite structural stiffness. This is because of the alternating structural stiffness of SMA under variable thermal loading due to twinning and detwinning behavior in a solid-state phase transformation. Some of the additional examples are presented to demonstrate the proposed approach's versatility and efficacy. In general, this research contributes to understanding smart material integration in laminated composites and offers a robust numerical framework for structural stiffness in diverse applications.

Keywords: ABAQUS, composites, FSDT, SMA, structural stiffness, smart material

1 Introduction

Laminated composites play an integral role in our daily lives, offering distinctive behavior and a wide range of applications tailored to specific needs. These materials are characterized by their unique properties, and researchers are dedicating substantial efforts to optimize their efficiency and cost-effectiveness across various sectors such as aerospace, marine, medical, and defense. In order to evaluate the structural stiff-

***Corresponding author: Dr. Erukala Kalyan Kumar**, Department of Mechanical Engineering, National Institute of Technology, Rourkela 769008, Odisha, India; Department of Mechanical Engineering, Presidency University, Itgalpur, Rajanakunte, Bengaluru 560064, Karnataka, India, e-mail: kalyan.me320@gmail.com, erukala.kalyan@presidencyuniversity.in

https://doi.org/10.1515/9783111515717-003

ness of these composite materials, several geometric parameters are evaluated such as deflection, natural frequency, buckling, and dynamic. Some of the studies related to natural frequency of laminated composites are: considering the frequency's impact on composite plates within a thermal environment, extensive research has been undertaken under various conditions, yielding results in accordance with requirements. Natural frequency analysis of multi-layered composite plates and shells is conducted using the curved triangular shell element [17]. A unique system for analyzing stiffened laminated plate frequencies was proposed based on state vector equation theory [14]. Free vibration analysis is performed on diverse structures, encompassing single and doubly curved laminate composite shells and plates with uninterrupted thickness variations. Numerical solutions are achieved through the generalized differential quadrature (GDQ) method. Theoretical approaches considering higher-order structural theories were presented [1]. Further, natural frequencies and buckling stresses of laminated plates were predicted using a generic spline finite strip capability [16] using FSDT) and classical plate theory. The investigation continues in the work of Lee and Han [11], delving into the eigenfrequencies of isotropic and composite laminates.

Modern research endeavors are directed towards enhancing laminated composites by integrating smart materials, thereby reinforcing the base structure. The integration of smart materials marks a pivotal advancement in the field, aimed at bolstering the capabilities of laminated composites. Among the assortment of smart materials employed for this purpose, shape memory alloys (SMAs), magnetorheological materials, and piezoelectric materials stand out prominently [2]. These materials contribute to the creation of intelligent composites capable of adapting to changing conditions, actively responding to stimuli, and enhancing overall performance. SMAs are more prominent reinforcement for structural applications including self-healing capabilities in the parent structure [9]. This progressive traditional laminated composites with cutting-edge smart materials hold substantial potential for revolutionizing applications across various industries, enabling innovative solutions to modern engineering challenges. The incorporation of SMAs presents a novel strategy that numerous researchers have adopted to enhance the integrity of the host structure. These SMAs have demonstrated a notable enhancement in the structural rigidity of the material. Notably, NiTi, a prominent member of the shape memory alloy family, finds widespread utilization in various structural applications. Now, the SMA-reinforced composites are gaining more attention due to their inherent applications. Some of the studies utilized SMA as reinforcement using higher-order shear deformation theory and FSDT studies are presented. Duan et al. [5] explored a finite element (FE) formulation and solution approach to understand the behavior of composite plates containing embedded SMA at elevated temperatures. Cho and Rhee [3] presented a nonlinear FE analysis method for modeling hybrid laminate composite shells that include embedded SMA wires, considering both structural and thermal loads simultaneously. Mahabadi et al. [13] provided an analytical closed-form solution for the free vibration response of a hybrid composite plate reinforced with SMA fibers. The analysis considered temperature-dependent mechanical properties of the plate. Using the FSDT, they derived the governing equations using Hamilton's principle and determined the fundamental natural frequency through the Ritz method at different temperatures. Kamarian and Shakeri [7] investigated the eigenfrequency behavior of laminated composite skew plates containing embedded SMA under thermal loads. They derived the governing equations using the FSDT and solved them using the GDQ technique, a precise and efficient numerical method. Some of the recent studies incorporated SMA as reinforcement to the damaged laminated composite structure and analyzed for free vibration [6], thermal buckling [10], and bending analysis [8] under various parameters.

Considerable research efforts have been directed towards exploring the natural frequency analysis of laminated composites. However, there has been a notable scarcity of studies focused on laminated composites embedded with SMAs in the same context. In this study, the finite strip-based FSDT combined with ABAQUS simulation software is employed to examine these responses under thermomechanical loading conditions. Moreover, this study presents various numerical examples of SMA-related parameters under different conditions, demonstrating a notable enhancement (up to 35%) in eigenresponses, transitioning from scenarios without SMA (V_f = 0%) to those incorporating SMA (V_f = 30%). The present work breaks new ground by examining laminated composites embedded with SMAs using advanced simulation tools, coupled thermomechanical analysis, and a diverse exploration of parameters. These advancements not only contribute to the understanding of these composites but also have the potential to impact a variety of materials science applications.

2 Methodology

The composite is modeled in a rectangular shape measuring 250 × 250 × 8 mm, comprising a total of eight layers, and this configuration was utilized for all the tests. For meshing, a linear pattern was adopted in the current modeling process, with an approximate global size of 0.05 mm. Furthermore, all sides of the square composite all sides are under a clamped condition ($U_x = U_y = U_z = \Theta_x = \Theta_y = \Theta_z = 0$), where U_x, U_y, U_z and Θ_x, Θ_y, Θ_z are the translational and rotational displacements in the x, y, and z axes, respectively. The stacking sequence employed is [0°/90°/90°/0°/0°/90°/90°/0°], and the analysis is carried out over a temperature range of up to 60 °C (with a reference temperature of 20 °C). Thermomechanical properties for epoxy-glass fiber [6, 12] and SMA (E_A = 67 GPa; E_M = 26.3 GPa, martensite temperatures: M_S = 18.4 °C; M_f = 9 °C, austinite temperatures: A_S = 34.5 °C; A_f = 49 °C, strain ε_L = 0.067, coefficient of thermal expansion $\alpha_S = 10.26 \times 10^{-6}$ 1/°C, and density (ρ) = 6,450 kg/m^3). The modulus of elasticity and recovery stress values for SMA are obtained from Cross et al. [4]. These structural, geometric characteristics, and material attributes are kept constant across the subsequent examples, unless explicitly stated otherwise. The simulation procedure encompasses a series of steps to attain the desired results. Beginning with the definition of the rectangular-laminated composite plate's geometry are specified. A meticulous meshing process ensues, ensuring a finely detailed mesh that accurately captures relevant characteristics. Material properties are assigned to individual layers of the composite plate. The subsequent phase involves the application of clamped and rigid boundary conditions across all sides of the composite plate, thereby simulating its securely fixed configuration. Further, post-processing steps are followed to obtain solution. The mathematical modeling of the current problem is as follows.

2.1 Theoretical Formulation

The current simulation model uses first-order mid-plane displacement kinematics [15]:

$$\begin{aligned} U_x(x,y,z) &= u_x(x,y,t) + z\phi_x(x,y,t) \\ V_y(x,y,z) &= v_y(x,y,t) + z\phi_y(x,y,t) \\ W_z(x,y,z) &= w_z(x,y,t) + z\phi_z(x,y,t) \end{aligned} \tag{1}$$

The U_x, V_y, and W_z vectors represent the displacements of a point within the shell at a given time t along orthogonal axes. On the other hand, the variables ϕ_x, ϕ_y, ϕ_z and u_x, v_y, w_z describe rotations and displacements of an arbitrary point located at the mid-plane.

2.2 Governing Equations

The equilibrium equation of motion for natural frequency responses is stated as

$$[M]\{\ddot{u}\} + [K]\{u\} = 0 \tag{2}$$

where $[M]$ and $[K]$ represent mass and stiffness matrix and $\{\ddot{u}\}$ and $\{u\}$ are acceleration and displacement vector, respectively. Additionally, eq. (2) has been solved to obtain the eigenvalue and its corresponding eigenvectors for the laminated structures as follows:

$$([K] - \omega^2[M])\{\Delta_o\} = 0 \tag{3}$$

Here, ω is the natural frequency and $\{\Delta_o\}$ represents eigenvector.

3 Results and Discussion

Moreover, the model's physical representation undergoes transformation into a simulation model via the FSDT method, incorporating the FEM for subsequent numerical assessments. The efficacy of this model is assessed through the analysis of frequency responses in laminated composites. To ascertain its precision, the numerical outcomes are compared with established reference data. The ensuing sections provide an in-depth examination of the computed results and the impact of input parameters on modal responses.

3.1 Convergence and Validation Study

Firstly, laminated composite embedded with SMA is analyzed for convergence study, with $[0/90/90/0]_S$, and it is computed for various mesh sizes (refer to Tab. 1). It is evident that the difference after 6 × 6 is minimal and it is opted for computing the additional illustrations. The geometrical parameters of laminate and SMA considered are provided in earlier section which is same as additional examples.

Further, a simulation model of a rectangular laminated composite plate measuring 0.6 × 0.4 × 0.005 m is modeled using ABAQUS simulation software for validation purpose. The composite plate consists of five layers with equivalent thicknesses. The laminate follows a stacking sequence of [0/90/0/90/0]. The material properties for each layer are as follows: E_1 = 132 GPa, E_2 = 10.3 GPa, G_{12} = 6.5 GPa, G_{13} = 6.5 GPa, G_{23} = 3.91 GPa, and Poisson's ratio (ν) = 0.3. The density (ρ) is 1,570 kg/m^3, while the coefficients of thermal expansion are $\alpha_1 = 1.2 \times 10^{-6}$/°C and $\alpha_2 = 2.4 \times 10^{-5}$/°C. Temperature conditions are meticulously set up, involving a temperature change (ΔT) that varies incrementally by 20 °C. This range commences from 0 °C and extends to 80 °C, all while maintaining an ambient or reference temperature of 0 °C. The chosen analysis type, often modal analysis, facilitates the extraction of the composite plate's natural frequencies under the defined temperature conditions. The simulation is executed for each distinct temperature configuration, yielding a comprehensive set of results. The natural frequency responses for various thermal conditions are validated with the available literature [12] as shown in Fig. 1.

3.2 Additional Examples

In addition, a thorough examination of the eigenfrequencies of the numerical model is conducted, incorporating various structural design parameters such as aspect ratio, thickness ratio, stacking sequence, and the positions of SMA layers to generate novel findings.

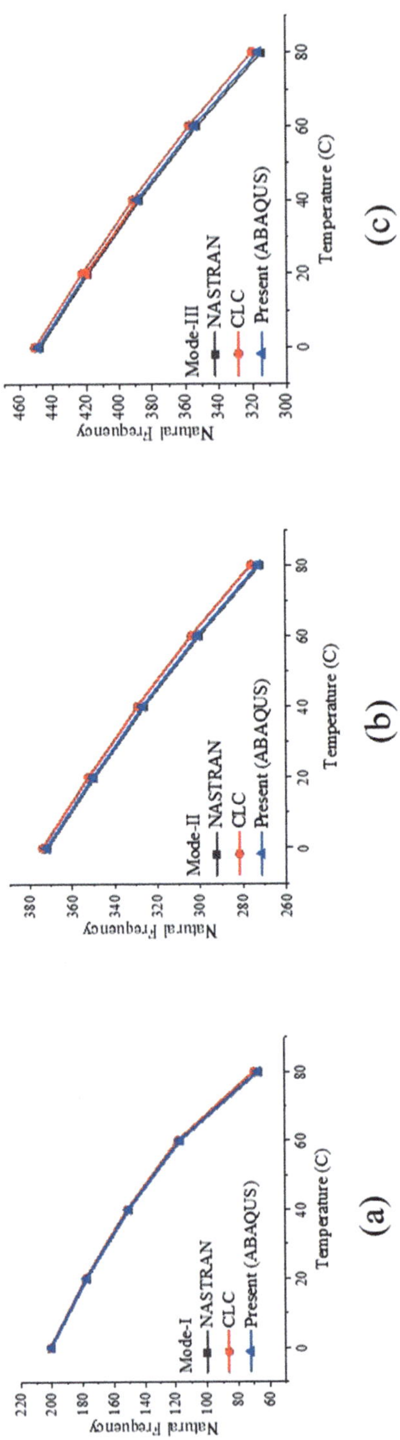

Fig. 1: Thermal frequency validation of laminated composite plate for all the modes: (a)–(c) modes I–III [12].

Tab. 1: Convergence study of laminated composite embedded with SMA.

Mesh size	1 × 1	2 × 2	3 × 3	4 × 4	5 × 5	6 × 6	7 × 7	8 × 8
Eigenvalue	71.323	67.722	66.199	65.477	65.076	64.833	64.675	64.566

3.2.1 Influence of Natural Frequency Responses for Different Aspect Ratios

The study involves an in-depth analysis of the natural frequency responses exhibited by laminated composite plates. This analysis incorporates a range of SMA volume fractions (V_f) from 0% to 30% and variations in aspect ratios ($b/a = 1.25–2$) under clamped boundary conditions as shown in Fig. 2(a). The natural frequencies of the plates progressively increase with increase in width to length ratio. This observed phenomenon can be attributed to the heightened stiffness exhibited by the plates as they become elongated relative to their width. Furthermore, the outcomes of this investigation underscore another intriguing finding: as the volume fraction of composite material within the plates increases, there is a corresponding uptrend in the natural frequency responses. The significant study revolves around the integration of SMAs into the parent structure. These SMAs possess distinctive mechanical characteristics, notably the capacity to revert to a predefined shape when exposed to specific stimuli such as temperature variations. Consequently, the integration of SMAs within the composite structure plays a role in augmenting its stiffness, which, in turn, influences the observed trends in natural frequency.

3.2.2 Influence of Natural Frequency Responses for Different Stacking Sequence

The analysis of eigenresponses in laminated composite plates, conducted across various temperature conditions, has yielded intriguing findings concerning the choice of layup sequences. A volume fraction of 25% for reinforcing SMA into laminated composites is likely chosen based on considerations such as achieving desired properties and balancing SMA. Specifically, the results indicate that the $[45°/90°/90°/–45°]_S$ layup sequence consistently delivers the most favorable natural frequency responses. On the other hand, the $[90°/–45°/45°/90°]_S$ layup sequence consistently produces the least natural frequency responses, regardless of the temperature variations as presented in Fig. 2(b). The orientation of fibers in the composite layers significantly affects the overall stiffness of the structure. The $[45°/–90°/90°/–45°]_S$ layup sequence (45° and –45°), which can enhance the plate's stiffness and resistance to bending. This increased stiffness contributes to higher natural frequencies. Conversely, the $[90°/–45°/45°/90°]_S$ layup sequence consists of plies oriented mainly in the 90° direction, which can result in lower stiffness and, consequently, lower natural frequencies.

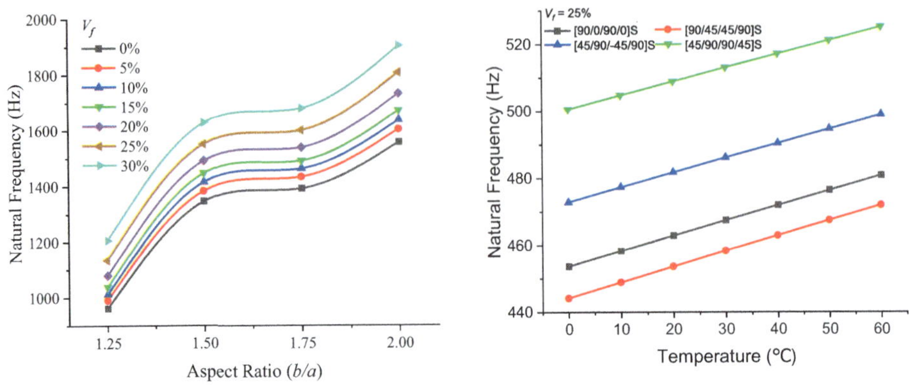

Fig. 2: Natural frequency responses of SMA embedded laminated composite plate: (a) aspect ratio and (b) stacking sequence.

3.2.3 Influence of Natural Frequency Responses for Different Thickness Ratio

In this section, an investigation was conducted on the eigenresponses of laminated composite plates with various thickness ratios (TR = 5–25) under different SMA temperatures (up to 60 °C) and clamped boundary conditions. The findings revealed that as the thickness ratio (length to thickness ratio) increases, the natural frequency responses tend to decrease as shown in Fig. 3. This decrease in natural frequency can be attributed to the thinning of the plate as the thickness ratio increases, resulting in a reduction of plate stiffness. This reduced thickness leads to a decrease in bending stiffness. Thinner plates are more flexible and have a lower resistance to bending and deformation. This increased flexibility allows the plate to vibrate more easily at lower frequencies, which contributes to the observed decrease in natural frequency responses. This trend holds true across all temperature conditions, with only minor improvements observed when varying temperatures from 0 °C to 60 °C. The accompanying Fig. 3 illustrates that lower thickness ratios lead to higher responses, while higher thickness ratios yield lower responses.

3.2.4 Influence of Natural Frequency Responses for Various SMA Layer Positions

In this section, we delve into the eigenresponses exhibited by laminated composite panels with embedded SMA layers, deliberately positioned across various layers (ranging from the first to the eighth layer). The analysis of the composite plate covers scenarios with both single and double layers of SMA, each subjected to varying SMA activation temperatures (elevated up to 60 °C). The volume fraction of SMA in each layer is controlled at 12.5%, and when using a double-layer configuration, it amounts

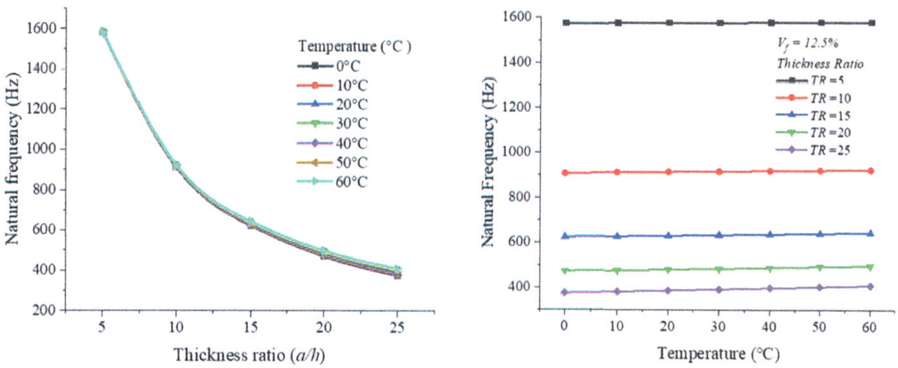

Fig. 3: Natural frequency responses of SMA embedded laminated composite plate for different thickness ratio and SMA temperatures.

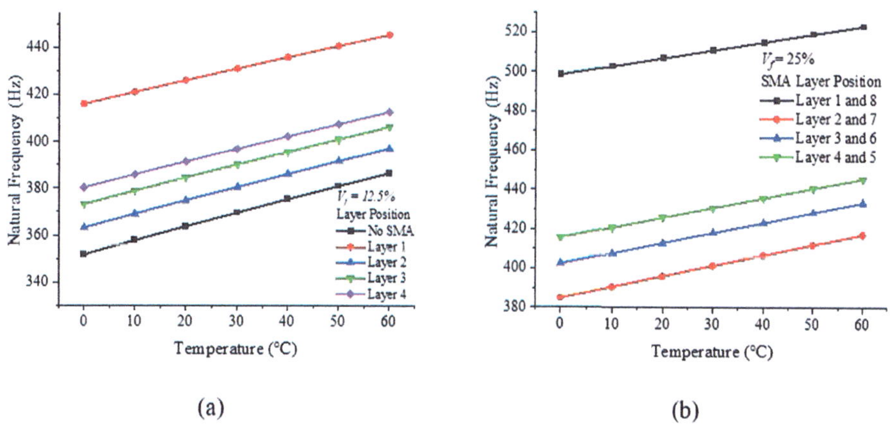

Fig. 4: Natural frequency responses of SMA embedded laminated composite plate: (a) single layer and (b) double layer.

to 25%. Therefore, the findings illuminate that placing the SMA in the first layer, where it has the most significant impact on stiffness, can lead to higher natural frequencies. Placing the SMA in the first and last layers creates a more balanced distribution of stiffness throughout the panel. This configuration ensures that the SMA's effects on stiffness are spread evenly across the structure, leading to better overall results in terms of natural frequencies. The activation temperature of SMAs plays a critical role in their behavior. By varying the activation temperature, you can control the extent to which the SMA contributes to the stiffness of the composite panel. When the activation temperature falls within the austenite finish temperature range, the SMA undergoes a phase transformation that significantly influences its stiffness. Increasing the temperature within this range makes the SMA more effective in enhancing the panel's stiffness, which in turn affects the natural frequencies. These factors collectively influence the stiffness and, consequently, the natural frequencies of the composite panel, leading to the observed trends and results in the analysis as shown in Fig. 4.

4 Conclusion

This study uses SMAs to enhance stability and stiffness in laminated composites through advanced modeling. Obtained outcomes confirm that SMA integration significantly improves natural frequency responses and structural stiffness. Some of the conclusions drawn from the study are as follows:

- Integration of SMAs significantly boosts natural frequency responses and structural stiffness. They are successful in improving laminates under various thermomechanical loads.
- The research contributes to the advancement of smart material integration in laminated composites. It offers a robust numerical framework applicable to diverse scenarios for enhancing structural stiffness.
- The presence of SMA volume fraction in laminates has influenced the eigenresponses positively across various geometric parameters, including aspect ratio, thickness ratio, and layup sequence and positions, and shown a significant improvement up to 35% in terms of eigenresponses.

The findings are showcased concerning SMA temperature and volume fractions, considering both geometric factors and SMA-related variables. This provides insight into the significance of these smart materials in enhancing structural characteristics.

References

[1] Bacciocchi M, Eisenberger M, Fantuzzi N, Tornabene F, Viola E. Vibration Analysis of Variable Thickness Plates and Shells by the Generalized Differential Quadrature Method, Compos. Struct. 2016;156:218–37.

[2] Bahl S, Nagar H, Singh I, Sehgal S. Smart Materials Types, Properties and Applications: A Review. Mater Today Proc. 2020;28:1302–06.

[3] Cho HK, Rhee J. Nonlinear Finite Element Analysis of Shape Memory Alloy (SMA) Wire Reinforced Hybrid Laminate Composite Shells. Int J Non Linear Mech. 2012;47:672–78.

[4] Cross WB, Anthony HK, Frederick JS. Nitinol Characterization Study, NASA. Langley Research Center; 1969.

[5] Duan B, Tawfik M, Goek SN, Ro -J-J, Mei C. Vibration of Laminated Composite Plates Embedded with Shape Memory Alloy at Elevated Temperatures. Smart Struct Mater 2000 Ind Commer Appl Smart Struct Technol. 2000;3991:366–76.

[6] Erukala KK, Mishra PK, Dewangan HC, Panda SK, Dwivedi M. Damaged Composite Structural Strength Enhancement under Elevated Thermal Environment Using Shape Memory Alloy Fiber. Acta Mech. 2022;233:3133–55.

[7] Kamarian S, Shakeri M. Natural Frequency Analysis of Composite Skew Plates with Embedded Shape Memory Alloys in Thermal Environment. AUT J Mech Eng. 2017;1:179–90.

[8] Kumar EK, Panda SK, Mahmoud SR, Balubaid M. Influence of Active SMA Fibre on Deflection Recovery Characteristics of Damaged Laminated Composite Theoretical and Experimental Analysis. Fibers Polym. 2023;24:3261–77.

[9] Kumar EK, Patel SS, Kumar V, Panda SK, Mahmoud SR, Balubaid M. State of Art Review on Applications and Mechanism of Self-Healing Materials and Structure. Arch Comput Methods Eng. 2023;30:1041–55.

[10] Kumar EK, Sharma N, Panda SK, Mahmoud SR. Numerical Prediction of Thermal Buckling Load Parameters of Damaged Polymeric Layered Composite Structure and Reversal of Strength Using SMA Fibre. Arch Appl Mech. 2022;92:3829–45.

[11] Lee WH, Han SC. Free and Forced Vibration Analysis of Laminated Composite Plates and Shells Using a 9-node Assumed Strain Shell Element. Comput Mech. 2006;39:41–58.

[12] Li X, Yu K, Han J, Song H, Zhao R. Buckling and Vibro-acoustic Response of the Clamped Composite Laminated Plate in Thermal Environment. Int J Mech Sci. 2016;119:370–82.

[13] Mahabadi RK, Shakeri M, Pazhooh MD. Free Vibration of Laminated Composite Plate with Shape Memory Alloy Fibers. Lat Am J Solids Struct. 2016;13:314–30.

[14] Qing G, Qiu J, Liu Y. Free Vibration Analysis of Stiffened Laminated Plates. Int J Solids Struct. 2006;43:1357–71.

[15] Szekrényes A. Analysis of Classical and First-order Shear Deformable Cracked Orthotropic Plates. J Compos Mater. 2014;48:1441–57.

[16] Tan D, Dawe DJ. Buckling and Vibration Analysis of Composite Laminated Plates and Shells Using General Spline Function. Compos Struct. 1997;40:25–42.

[17] Zhu J. Free Vibration Analysis of Multilayered Composite Plates and Shells with the Natural Approach. Comput Methods Appl. Mech. Eng. 1996;130:133–49.

Shubhra Shitole, S. Vincent, Abdel-Hamid Ismail Mourad*,
and Priya Sudha

Modelling Techniques to Analyse the Functionality of Metallic Glass Alloys as Biomedical Implants: A Mini-review

Abstract: Metallic glass alloys have emerged as a promising class of materials for orthopaedic knee implants due to their unique combination of mechanical properties, including high strength, excellent wear resistance, and biocompatibility. Understanding the mechanical behaviour and performance of metallic glass under physiological stress circumstances is essential for their design and optimization. The modelling methods used to examine the mechanical reaction and long-term functionality of metallic glass biomedical applications are reviewed in this chapter. We explore in depth several modelling techniques, such as finite element analysis, molecular dynamic simulations, statistical modelling, and machine learning. The review addresses issues such as stress distribution prediction, wear behaviour, and assessment of fatigue life. Also discussed are the drawbacks and difficulties of modelling these intricate materials. This review intends to promote the development of advanced metallic glass with improved mechanical properties, increased lifespan, and improved patient outcomes by offering a thorough examination of modelling approaches.

Keywords: Metallic glass alloys, biocompatibility, finite element analysis, molecular dynamic simulations and statistical modelling

1 Introduction

In recent decades, the quest for improved materials for biomedical implants has been at the forefront of materials science research. The importance of metallic glass (MG)

*Corresponding authors: **Abdel-Hamid Ismail Mourad**, Mechanical and Aerospace Engineering Department, College of Engineering, United Arab Emirates University, Al-Ain 15551, United Arab Emirates; National Water and Energy Center, United Arab Emirates University, Al Ain 15551, United Arab Emirates, e-mail: ahmourad@uaeu.ac.ae
Shubhra Shitole, Mechanical and Aerospace Engineering Department, College of Engineering, United Arab Emirates University, Al-Ain 15551, United Arab Emirates, e-mail: 202090206@uaeu.ac.ae
S. Vincent, Department of Mechanical Engineering, BITS Pilani, Dubai Campus, Dubai 345055, United Arab Emirates, e-mail: vincent@dubai.bits-pilani.ac.in
Priya Sudha, Department of Mechanical Engineering, BITS Pilani, Dubai Campus, Dubai 345055, United Arab Emirates, e-mail: priyasudha32@gmail.com

https://doi.org/10.1515/9783111515717-004

alloys in biomedical applications lies in their exceptional mechanical properties, corrosion resistance, and biocompatibility, which make them promising candidates for implants [1]. Their potential to enhance implant longevity and performance underscores their significance in biomedical engineering. MG alloys offer opportunities for designing tailored implants to meet specific patient needs, driving advancements in implant technology (orthopaedic and dental implants to cardiovascular stents) and improving patient outcomes [2–5].

Designing and refining MG alloys for biomedical use is challenging due to their complex microstructures and composition-dependent characteristics. Advanced modelling tools have facilitated a better understanding and manipulation of these features, requiring a thorough exploration of the atomic-scale behaviour of these alloys [6]. For instance, Babilas et al. [7] utilized reverse Monte Carlo modelling to elucidate the local atomic structure and propose an atomic structural model to analyse a quaternary Mg-based MG alloy. Finite element modelling (FEM) was implemented by Liu et al. [8] to study mechanical properties such as deformation behaviour. Another modelling technique, molecular modelling simulations, was demonstrated by Chen et al. [9], to reveal the fractal nature of MG structure, indicating a dimensionality crossover within these alloys.

As machine learning (ML) and materials informatics gain prominence, they offer a comprehensive approach to studying materials by integrating insights from experimentation, theory, and simulations [10–12]. To advance the development and utilization of MG alloys for biomedical implants, this review paper aims to comprehensively investigate the pivotal role played by modelling techniques. By navigating the complex realm of computational tools, simulations, and modelling approaches, we aim to shed light on the challenges and possibilities in this fast-paced industry, ultimately paving the way for safer, more effective biomedical implants. A schematic outline for the objective of this paper is depicted in Fig. 1.

The study of MG alloys employs various modelling techniques, each providing valuable insights into material behaviour. From ML predicting properties to physics-based simulations elucidating atomic interactions, these methods collectively advance our understanding. Experimental testing validates findings, while empirical relationships and statistical models offer quick estimations. Finite element analysis (FEA) examines mechanical behaviour, while molecular dynamics (MD) scrutinizes atomic dynamics. Integrating these approaches yields a comprehensive understanding, overcoming individual limitations. Table 1 gives a general overview of the four modelling techniques covered in this chapter.

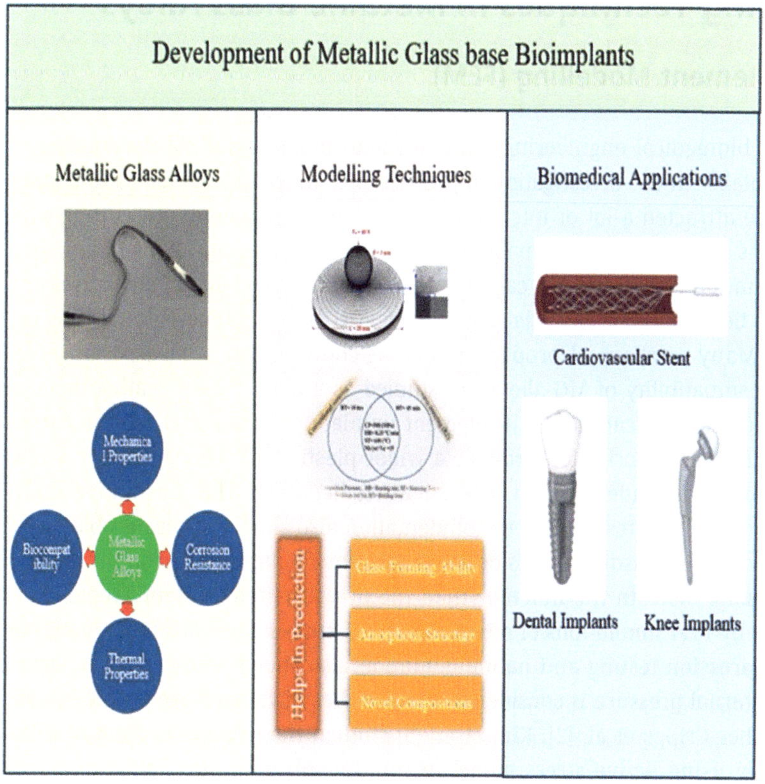

Fig. 1: A schematic representation of modelling techniques used to analyse metallic glass alloys.

Tab. 1: Modelling techniques generally used to analyse metallic glass alloys and their shortcomings.

Models	Applications	Limitations and challenges
Finite element analysis (FEA)	Analysing mechanical behaviour, stress distribution, and deformation	Requires expertise in numerical methods and computational resources.
Molecular dynamic (MD) modelling	Analysing atomic-scale behaviour and structural dynamics	Requires significant computational resources and expertise in molecular simulations. They may be limited to small systems and short timescales.
Statistical modelling	Analysing correlations and trends in data predicting material properties based on statistical methods	Relies on assumptions about data distributions and may struggle with nonlinearities.
Machine learning	Predicting material properties such as glass-forming ability (GFA), mechanical properties and phase transitions	Requires large datasets for training and may struggle with complex relationships or limited data availability.

2 Modelling Techniques in Metallic Glass Alloys

2.1 Finite Element Modelling (FEM)

In the field of biomedical engineering, finite element modelling (FEM) has become a crucial tool, notably in the investigation and use of MG alloys for biomedical implants. MG alloys have attracted a lot of interest as prospective components for cutting-edge implant designs because of their enviable mechanical qualities and biocompatibility. FEM, a computational method that can simulate and analyse complex structural and biomechanical behaviours, is essential for utilizing MG alloys' full potential in this important area. Many mechanical properties such as compression, wear resistance, fatigue, and biocompatibility of MG alloys are studied as a part of investigating them as probable biomedical applications. Finite element simulations of nanoindentation curves using the Mohr-Coulomb criterion produce a wider plastic zone beneath the indenter than the pressure-independent Tresca criterion as studied in [1]. The constitutive analysis based on compression testing and nanoindentation also shows a considerable divergence when internal pressure is considered. The Mohr-Coulomb criterion produces a wider plastic zone beneath the indenter than the pressure-independent Tresca criterion, as shown by FEA simulations of nanoindentation curves. The constitutive analysis based on compression testing and nanoindentation also shows a considerable divergence when internal pressure is considered. Such studies have been previously investigated by Frischer-Cripps et al. [2]. Kim et al. [3] studied the stress-overshoot of a Pd-based MG alloy using fictive stress model in conjunction with Maxwell viscoelastic model under uniaxial compression. The deformation studies are in good concurrence with the experimental results. The fictive stress model describes the relaxation behaviour of amorphous materials like MG alloys. The model is based on the concept of fictive stress which represents the internal stresses within the material which arise due to the structural changes occurring during rapid cooling or deformation. This model utilizes empirical relationships to characterize the evolution of fictive stresses over time or temperature. The Maxwell viscoelastic model is a classical model consisting of a spring representing the elastic behaviour and a dashpot representing the viscous behaviour, which describes the time-dependent behaviour of materials. The equation governing this model is given below and the system can be picturized as in Fig. 2.

$$\sigma(t) = \sigma_0{}^{*} e^{-\frac{t}{\tau}} \tag{1}$$

where $\sigma(t)$ is the stress at time t, σ_0 is the initial stress, and τ is the characteristic relaxation time.

The fatigue behaviour of a structural material substrate coated with thin-film MG (TFMG) alloys was investigated in the study by Jia et al. [4]. The Rudnicki-Rice instability theory and the free volume model, which were implemented using FEM, were used to study the initiation and propagation of shear bands under bending deforma-

Fig. 2: Schematic representation of classical Maxwell viscoelastic model.

tion. The radial shear band patterns under indentation for both the bare substrate and the coated substrate were revealed when the Rudnicki-Rice theory was used to predict the directional growth of shear bands. The FEM results matched the outcomes of the experiments. Similar research was also done in [5, 6, 8, 13], highlighting the observed delay in shear band initiation and propagation that was related to the increased ductility of TFMG alloys. This improved ductility enables efficient stress absorption and even stress distribution across the underlying substrate. The tests were expanded to investigate the fatigue behaviour of bulk MG (BMG) substrates coated with TFMG, further supporting the contribution of improved ductility in postponing shear banding events.

With a focus on spherical indentation, Yu et al. [14] used FEM to examine the fatigue resistance of a Zr-based BMG alloy coated with a TFMG. In the TFMG-coated samples, the FEM analysis showed that shear-band nucleation and propagation were significantly delayed. Notably, the TFMG-coated BMG's endurance limit showed a startling 33% increase in comparison to the BMG substrate alone. Cycled bending tests on uncoated and TFMG-coated BMG substrates were also used in the study, along with FEM for fatigue analysis. These studies considered complex geometry and loading patterns, isotropic elastic-plastic constitutive laws, and shear offset effects. Utilizing an indentation model that implemented the free volume model suggested in [10], it was determined whether there was a delay in the creation and propagation of shear bands. The tests were carried out using Dassault Systèmes Simulia Corp.'s Abaqus finite element program, version 6.13, in Providence, Rhode Island, USA. While the presented research has concentrated on a small selection of MG alloys, there is an opportunity to extend to a larger selection of MG alloys to acquire a thorough understanding of these properties over a wide range of materials.

Wear resistance is another pre-requisite property for materials being considered for biomedical applications. Many researchers used FEM to study wear resistance, in addition to the MG alloys' fatigue resistance. Hua et al. [11] used the Hertzian calculation along with FEM to determine the good wear resistance of Zr, Ti, Mg, and Fe-based MG alloy. Figure 3 shows the depiction of the FEM implemented for the analysis where the ceramic sphere is shown to generate the dent under the application of F_n force. The wear resistance depends on the coupling material and the base material [12].

Abaqus™ was used to create the FEM axisymmetric representation of the Hertzian contact mechanics between the Fe, Zr, Ti, and Mg-based MG alloy and corresponding coupling pair of either Al_2O_3, Si_3N_4, and ZrO_2. The Fe-based MG alloy showed

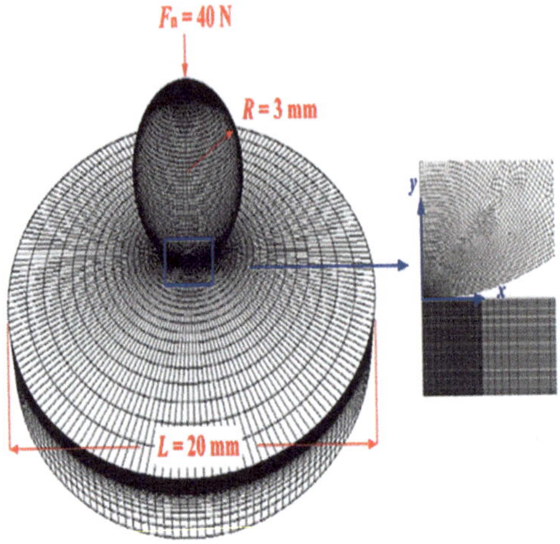

Fig. 3: Finite element model used for analysing the dry wear performance of metallic glass alloy and coupling material [11].

better corrosion resistance as compared to the other tested alloys. The coupling material is chosen based on the material being used in orthopaedic implants. It was concluded that Fe-based MG alloy showed the highest wear resistance when coupled with Si_3N_4 which was attributed to the ultrahigh fracture strength of the alloy.

Huang et al. [15] studied the viability of Zr-based MG alloy as a candidate for stent application. Abaqus (standard finite element analysis (FEA) package) was implemented to study the stress-strain fields of the stent. A stent vessel system was created to maintain an allowance of 0.1 mm between the stent and the vessel. Mooney-Rivlin hyperplastic material was designated to simulate the vessel of the stent vessel system [16]. Von misses' stresses and maximum principal strain were applied on the fully expanded stent model. The Von misses' stresses were significantly lower than the yield stress of the material, thus emphasizing the bearing of sustaining physiological loadings. The highest strains on both types of stents were substantially lower than their elastic limits, suggesting that the blood vessel and stent deformation would be contained within the elastic zone. The stents made from Zr-based MG alloys are more flexible and deform easily with the blood vessel during beats and thus more compliant as stent materials. Other studies on the mechanical properties of Zr-based MG alloys as stents have been discussed in [17] and the strain conditions during expansion and after deployment are studied in [18].

2.2 Molecular Dynamic (MD) Simulations

To study the dynamic behaviour of molecular systems over time, many scientific disciplines use MD simulations, a potent computer approach. MD simulations are particularly helpful in the fields of materials science, chemistry, and biology for revealing the minute details of atomic and molecular interactions. These simulations simulate the motion and interactions of particles inside a specified system using Newtonian dynamics, offering insights on thermodynamic characteristics, structural alterations, and dynamical processes. Different kinds of MD simulations are available to meet different research goals. Figure 4 illustrates the different types of MD simulation models prominently used in studies. Understanding and developing MG alloys for biomedical implants rely heavily on MD simulations. These simulations make use of computational models to examine the atomic-scale dynamics and behaviour of MGs, revealing details about their mechanical characteristics, interactions with biological surroundings, and structural stability. MD simulations enable researchers to anticipate and optimize the performance of MG alloys in the context of biomedical implants, where characteristics such as biocompatibility, corrosion resistance, and mechanical strength are crucial. These simulations assist in customizing alloy compositions to promote biocompatibility and prevent corrosion, two factors that are essential for implant longevity and long-

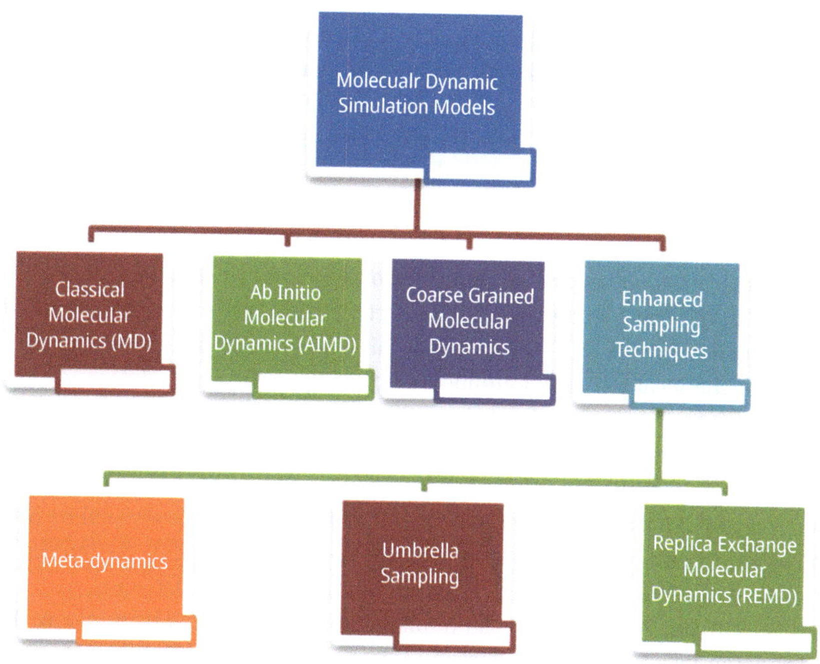

Fig. 4: Classification of different types of molecular dynamic (MD) simulation techniques.

term success. These interactions between the MG surface and biological components are complicated, and they are virtually mimicked by these simulations. Additionally, MD simulations make it easier to investigate deformation mechanisms, which in turn aids in the development of MG implants with enhanced mechanical characteristics that can endure the dynamic environments found inside the human body. The creation of safer, more reliable, and functionally superior materials for medical purposes is greatly aided using MD simulations into the research of MG alloys as biomedical implants.

2.2.1 Characterization Studies

The embedded atom method (EAM) potentials, Honeycutt-Andersen indices, Voronoi tessellation method, radial distribution function, and coordinate numbers are frequently used in studies using MD simulations to examine the glass-forming process and atomic structure of MG alloys. The following is a list of debates based on these criteria.

Interatomic interactions are described by certain potential functions necessary for MD simulations. Studies [19–22] have used EAM potentials generated using ab initio molecular dynamics (AIMD) to explain the interatomic interactions of Pd-Si-based MG alloys. The same dataset was implemented by Celtek et al. [23] to analyse the interatomic interactions of another Pd-Si MG alloy to generate results which were in congruence with the finding from [19–22]. MD simulations were performed using DL_POLY 2.0 simulation package. The parameters like structural functions $g(r)$ and $S(q)$ relating information about glass formation process of the systems and the evolution of short-range ordering were referred from [24–27]. The glass formation temperature (T_g) is an important parameter used to analyse the glass formation process. This was determined using mean work per atom (MWA) parameter. From the simulation it was concluded that with increasing value of Si, the T_g values decrease linearly. These results were in congruence with other studies conducted on similar Pd-Si MG alloys [28, 29].

Celtek et al. [27] have studied the effect of cooling rate on the atomic structure and glass formation process of Co-Zr-based MG alloy using EAM potentials. The characterization of the local structural atomic orders of the system was determined by Honeycutt-Andersen indices and Voronoi tessellation method. With a decrease in cooling rate there was a linear decrease in the glass forming temperature (T_g). The amorphous structure of the alloy is influenced by the presence of high icosahedral and face cantered cubic short-range order, which diminishes with decreasing cooling rate. The study confirmed the use of EAM interatomic potentials to study the impact of cooling rates on the structural and glass formation process of Co-Zr MG alloy.

Classical MD simulations were also implemented by Celtek et al. [30] using the same EAM potentials, Honeycutt-Anderson indices and Voronoi tessellation method to study the temperature effect on the atomic structure and glass forming system of

Zr-Cu-Al ternary MG alloy. It was concluded that the percentage of Al in the ternary alloy determines the glass forming ability of the ternary system with a maximum 20% of Al generating consistent results with literature [31–33].

2.2.2 Mechanical Properties

Apart from FEM, MD simulations can also be used to analyse mechanical properties as shear banding and plastic deformations of MG alloys. Shimizu et al. [34] implemented the aged-rejuvenated-glue-liquid (ARGL) shear band model which shows that the shear band propagation depends on value of applied stress with respect to the glue traction. The shear band will only propagate when the stress applied is greater than the glue traction. This result was proven using large-scale MD simulations.

Ye et al. [35] used MD simulations to study the effect of cyclic loading on the atomistic mechanism of Cu-Zr binary MG alloy. They concluded that cyclic loading causes atomic bonds to break. Notably, elastic softening at lower stress rates is mostly ascribed to the activation of bigger, more extended string-like atomic clusters that behave as "liquid-like" materials, indicating a viscoelastic response. Given that MG yielding is strain-controlled and correlated with variables like the elastic modulus or glass transition temperature (T_g), the continual elastic softening that is seen during cyclic loading may eventually result in yielding or shear banding.

Wang et al. [36] studied the impact of strain rate on shear band propagation and also the dependence on stress concentrations for MG alloys using Zr-based MG alloy. The large-scale atomic/molecular massively parallel simulator [37] was implemented to analyse the alloy. The dependence of stress concentrations on the strain rate sensitivity (SRS), which determined the propagation of shear bands is tested. It is established that if the alloy has higher stress concentrations, the corresponding SRS will be negative, which leads to shear band propagation. On the other hand, a positive SRS is noted with lesser or no stress concentrations. These results are in congruence with the literature [38–41].

2.3 Statistical Modelling

Exploration of MG alloys for biomedical applications requires multidimensional material analysis combining state-of-the-art materials science and statistical modelling methods. Creating new biomaterials with better mechanical properties, biocompatibility, and durability is a research goal, and statistical modelling is becoming an essential tool for its systematic investigation. Scientists can decipher complex correlations between alloy compositions, processing variables, and the resulting material qualities by using a variety of statistical tools, including models such as regression analysis, multivariate analysis, and optimization techniques. The material behaviour may be predicted, the best alloy combinations can be found, and the biocompatibility can be thoroughly evaluated thanks to this meticulous investigation. In the complex process of comprehending, improving, and customizing MG alloys for use in the field of biomedical engineering, statistical modelling acts as a compass. Using microwave sintering to fine-tune the material characteristics of an Mg-based MG alloy designed for use as a biomedical implant, Kumar et al. [42] developed a design of experiments (DOE) technique for their investigation. Due to its inherent benefits over spark plasma sintering and traditional sintering methods, microwave sintering was chosen. The DOE method made it possible to optimize important process variables that were applicable to both traditional and microwave sintering techniques. Compaction pressure, heating rate, sintering temperature, Niobium (Nb) concentration (wt.%), and holding duration were among the factors. Figure 5 shows the depicted values of these improved parameters, indicating a methodical and specific methodology to improve the material properties necessary for the envisioned biomedical implant application. The composition of the alloying elements was chosen as they have shown definite properties for orthopaedic implants [43].

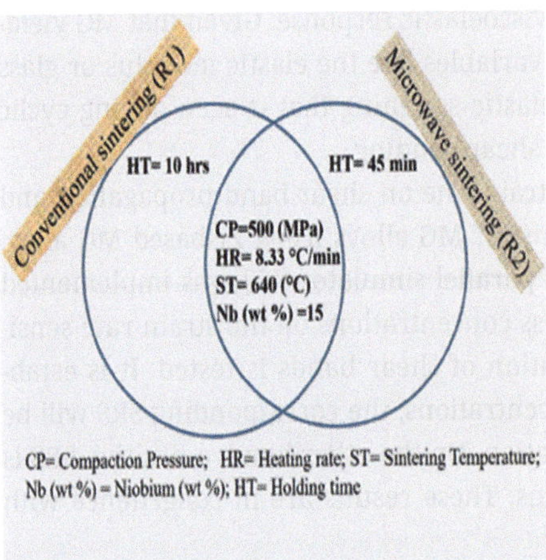

Fig. 5: Optimized process parameters used to manufacture the Mg-based MG alloys using developed DOE [42].

Bajpai et al. [44] employed statistical modelling to develop new high-entropy bulk MG alloys (HE-BMG) incorporating key thermodynamical and topological aspects of glass formation of the alloy. The parameters of enthalpy of chemical mixing (ΔH_{mix}), configurational entropy (ΔS_{config}) and entropy due to atomic size mismatch (ΔS_{σ}) are integrated into a single parameter using P_{HSS} model. Totally, 69 elements were used to determine this P_{HSS} parameter to predict different compositions for HE-BMG alloys.

2.4 Machine Learning (ML)

ML models have been extensively utilized to predict the GFA of MG. Various ML models, such as support vector regression (SVR), deep neural networks, and ensemble learning models, have been developed to predict critical parameters like the maximum casting diameter (D_{max}) [45–47]. These models address challenges like data imbalance and low prediction accuracy, enhancing the generalization capability and accuracy of predictions [48, 49]. ML approaches have also been applied to understand the factors influencing the GFA of MG alloys, aiming to guide alloy design and accelerate the discovery of novel glass formers. By leveraging ML techniques, researchers can efficiently predict and identify MG compositions with superior GFA, facilitating the development of advanced MG alloy systems.

Ward et al. [50] implemented ML along with statistical modelling to develop a BMG sample with a desirable property of supercooled liquid range (ΔT_x). The other desired properties analysed where ability to maintain amorphous state and critical casting diameter. The main steps involved in generating the ML model is: (i) collect – digital data which will be the foundation of the ML model; (ii) process – processing the collected data to develop training sets with defined inputs and outputs; (iii) represent – the processed data in form compatible with ML; and (iv) learn – using ML algorithms to learn the predictive model from training data. Figure 6 shows in detail schematization of these steps. The number of possible combinations analysed was about 2.7 million.

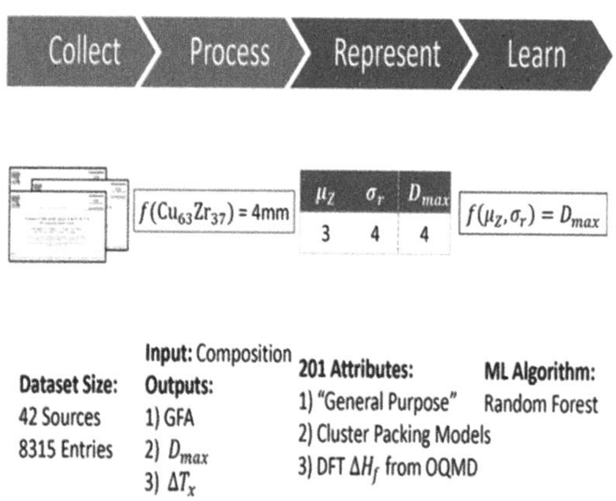

Fig. 6: Schematic representation of steps involved in generating a machine learning model [50].

ML techniques are instrumental in analysing MG alloys, enabling the extraction of critical correlations between composition, processing parameters, and material properties. ML is often used in combination with other modelling methods rather than as a standalone approach for analysing MG alloys. While ML techniques excel at extracting patterns from data and predicting material properties, they are typically integrated into broader computational frameworks that incorporate physics-based modelling, experimental data, and empirical relationships. This combined approach leverages the strengths of different methods, allowing for a more comprehensive understanding of MG behaviour and enabling more accurate predictions and optimizations.

3 Future Scope

Future uses of FEM, MD simulations, and statistical modelling for the analysis of MG alloys in biomedical applications have enormous potential, especially in addressing the crucial issue of biocompatibility. The developing environment underlines the necessity to investigate the biocompatibility characteristics of MG alloys, despite the fact that the present uses of these modelling tools have mostly concentrated on mechanical and thermal aspects. In order to create implants with specialized mechanical properties that are compatible with the biomechanics of the human body, FEM can be useful in predicting the mechanical response of MG under physiological settings. When used to explore the interaction between MG surfaces and living things, MD simulations can shed important light on the bioactivity and biocompatibility of these substances. The systematic optimization of MG for biomedical implants can be facilitated by statistical modelling, which can also play a key role in linking alloy compositions, processing variables, and structural aspects with biocompatibility attributes.

Additionally, the use of these modelling techniques can make a substantial contribution to the creation of novel biomedical implants and devices with improved biocompatibility and functionality. The multi-scale nature enables scientists to bridge the gap between atomic-level interactions, material characteristics, and macroscopic behaviour under physiological circumstances. Researchers can increase our understanding of how MG alloys function inside the human body by combining data on surface interactions, corrosion resistance, and the release of ions or nanoparticles into the biological milieu. The growing need for biomaterials that not only have mechanical integrity but also promote positive biological responses is met by this all-encompassing strategy. The growing need for biomaterials that not only have mechanical integrity but also promote positive biological responses is met by this all-encompassing strategy. Future MG alloys that excel in both structural performance and biocompatibility for a variety of biomedical applications will undoubtedly be designed thanks to the synergy of FEM, MD simulations, and statistical modelling as the field develops.

4 Conclusion

This succinct study examines various modelling approaches used to analyse MG alloys, including statistical modelling, ML, FEM, and MD simulation. Due to the difficulty in studying mechanical properties when using TFMG alloys, the investigation becomes especially pertinent when dealing with these materials. Utilizing these modelling methods becomes essential to avoid spending time and money on creating BMG alloys for investigation. The studies under discussion dive into several features of the mechanical characteristics of MG alloys, such as the initiation and propagation of shear bands, the endurance of fatigue, the behaviour of stress-strain, and plasticity. Examinations also cover basic characteristics such atomic structure, glass formation, and glass transition temperature (T_g), all of which are essential for structural applications in industry. The investigation of these qualities for biomedical applications, where biocompatibility, cytotoxicity, and antibacterial qualities are prioritized, is equally important. The studies discussed here set the foundations for additional research using these modelling tools, offering a thorough foundation for a more comprehensive knowledge of MG alloys in a variety of applications.

References

[1] Fornell J, Concustell A, Suriñach S, Li WH, Cuadrado N, Gebert A, et al. Yielding and Intrinsic Plasticity of Ti–Zr–Ni–Cu–Be Bulk Metallic Glass. Int J Plast. 2009;25:1540–59.

[2] Fischer-Cripps AC, Nicholson DW. Nanoindentation. Mechanical Engineering Series. Appl Mech Rev. 2004;57:B12–B12.

[3] Kim HS, Kato H, Inoue A, Chen H-S. Finite Element Analysis of Compressive Deformation of Bulk Metallic Glasses. Acta Mater. 2004;52:3813–23.

[4] Jia H, Liu F, An Z, Li W, Wang G, Chu JP, et al. Thin-film Metallic Glasses for Substrate Fatigue-property Improvements. Thin Solid Films. 2014;561:2–27.

[5] Chu JP, Greene JE, Jang JSC, Huang JC, Shen Y-L, Liaw PK, et al. Bendable Bulk Metallic Glass: Effects of a Thin, Adhesive, Strong, and Ductile Coating. Acta Mater. 2012;60:3226–38.

[6] Chang YZ, Tsai PH, Jb L, Hc L, Jang JSC, Li C, et al. Zr-based Metallic Glass Thin Film Coating for Fatigue-properties Improvement of 7075-T6 Aluminum Alloy. Thin Solid Films. 2013;544:331–34.

[7] Babilas R, Łukowiec D, Temleitner L. Atomic Structure of Mg-based Metallic Glass Investigated with Neutron Diffraction, Reverse Monte Carlo Modeling and Electron Microscopy. Beilstein J Nanotechnol. 2017;8:1174–82.

[8] Chiang CL, Chu JP, Liu FX, Liaw PK, Buchanan RA. A 200nm Thick Glass-forming Metallic Film for Fatigue-property Enhancements. Appl Phys Lett. 2006;88.

[9] Cheng YQ, Ma E. Atomic-level Structure and Structure–property Relationship in Metallic Glasses. Prog Mater Sci. 2011;56:379–473.

[10] Spaepen F. A Microscopic Mechanism for Steady State Inhomogeneous Flow in Metallic Glasses. Acta Metallurgica. 1977;25:407–15.

[11] Hua N, Zhang X, Liao Z, Hong X, Guo Q, Huang Y, et al. Dry Wear Behavior and Mechanism of a Fe-based Bulk Metallic Glass: Description by Hertzian Contact Calculation and Finite-element Method Simulation. J Non Cryst Solids. 2020;543:120065.

[12] Hua N, Chen W, Wang W, Lu H, Ye X, Li G, et al. Tribological Behavior of a Ni-free Zr-based Bulk Metallic Glass with Potential for Biomedical Applications. Mater Sci Eng C. 2016;66:268–77.

[13] Lee CM, Chu JP, Chang WZ, Lee JW, Jang JSC, Liaw PK. Fatigue Property Improvements of Ti–6Al–4V by Thin Film Coatings of Metallic Glass and TiN: A Comparison Study. Thin Solid Films. 2014;561:33–37.

[14] Yu -C-C, Chu JP, Jia H, Shen Y-L, Gao Y, Liaw PK, et al. Influence of Thin-film Metallic Glass Coating on Fatigue Behavior of Bulk Metallic Glass: Experiments and Finite Element Modeling. Mater Sci Eng C. 2017;692:146–55.

[15] Huang L, Pu C, Fisher RK, Mountain DJH, Gao Y, Liaw PK, et al. A Zr-based Bulk Metallic Glass for Future Stent Applications: Materials Properties, Finite Element Modeling, and in Vitro Human Vascular Cell Response. Acta Biomater. 2015;25:356–68.

[16] Lally C, Dolan F, Prendergast PJ. Cardiovascular Stent Design and Vessel Stresses: A Finite Element Analysis. J Biomech. 2005;38:1574–81.

[17] Kastrati A, Dirschinger J, Boekstegers P, Elezi S, Schühlen H, Pache J, et al. Influence of Stent Design on 1-year Outcome after Coronary Stent Placement: A Randomized Comparison of Five Stent Types in 1,147 Unselected Patients. Cathet Cardiovas Interv. 2000;50:290–97.

[18] Kleinstreuer C, Li Z, Basciano CA, Seelecke S, Farber MA. Computational Mechanics of Nitinol Stent Grafts. J Biomech. 2008;41:2370–78.

[19] Daw MS, Baskes MI. Embedded-atom Method: Derivation and Application to Impurities, Surfaces, and Other Defects in Metals. Phys Rev B. 1984;29:6443.

[20] Sheng HW. No Title. EAM Potential Website; n.d.

[21] Guder V, Sengul S, Celtek M, Domekeli U. Pressure Dependent Evolution of Microstructures in Pd80Si20 Bulk Metallic Glass. J Non Cryst Solids. 2022;576:121290.

[22] Deng YH, Wen DD, Li Y, Liu J, Peng P. Local Atomic Structures of Amorphous Pd80Si20 Alloys and Their Configuration Heredity in the Rapid Solidification. Philos Mag. 2018;98:2861–77.

[23] Celtek M, Sengul S, Domekeli U, Guder V. Molecular Dynamics Simulations of Glass Formation, Structural Evolution and Diffusivity of the Pd-Si Alloys during the Rapid Solidification Process. J Mol Liq. 2023;372:121163.

[24] Celik FA, Korkmaz ET. Molecular Dynamic Investigation of the Effect of Atomic Polyhedrons on Crystallization Mechanism for Cu-based Cu-Pd and Cu-Pt Alloys. J Mol Liq. 2020;314:113636.

[25] Kazanc S, Celik FA, Yildiz AK, Ozgen S. Pressure Effect on Intermediate Structures during Transition from Amorphous to Crystalline States of Copper. Comput Mater Sci. 2007;40:179–85.

[26] Yildiz AK, Celik FA. Atomic Concentration Effect on Thermal Properties during Solidification of Pt-Rh Alloy: A Molecular Dynamics Simulation. J Cryst Growth. 2017;463:194–200.

[27] Celtek M, Şengül S. Effects of Cooling Rate on the Atomic Structure and Glass Formation Process of Co90 Zr10 Metallic Glass Investigated by Molecular Dynamics Simulations. Turk J Phys. 2019;43:11–25.

[28] Ballone P, Rubini S. Embedded-atom Model of Glass-forming Si-metal Alloys. Phys Rev B. 1995;51:14962.

[29] Chen N, Li Y, Yao K-F. Thermal Stability and Fragility of Pd–Si Binary Bulk Metallic Glasses. J Alloys Compd. 2010;504:S211–4.

[30] Celtek M, Sengul S, Domekeli U. Glass Formation and Structural Properties of Zr50Cu50-xAlx Bulk Metallic Glasses Investigated by Molecular Dynamics Simulations. Intermetallics (Barking). 2017;84:62–73.

[31] Ward L, Agrawal A, Flores KM, Windl W Rapid Production of Accurate Embedded-atom Method Potentials for Metal Alloys. ArXiv Preprint ArXiv:12090619 2012.

[32] Zhou XW, Johnson RA, Wadley HNG. Misfit-energy-increasing Dislocations in Vapor-deposited CoFe/NiFe Multilayers. Phys Rev B. 2004;69:144113.

[33] Cheng YQ, Ma E, Sheng HW. Atomic Level Structure in Multicomponent Bulk Metallic Glass. Phys Rev Lett. 2009;102:245501.
[34] Shimizu F, Ogata S, Li J. Theory of Shear Banding in Metallic Glasses and Molecular Dynamics Calculations. Mater Trans. 2007;48:2923–27.
[35] Ye YF, Wang S, Fan J, Liu CT, Yang Y. Atomistic Mechanism of Elastic Softening in Metallic Glass under Cyclic Loading Revealed by Molecular Dynamics Simulations. Intermetallics (Barking). 2016;68:5–10.
[36] Wang H, Zhao X, Tian B, Han Z, Yang P, Chen X, et al. Strain Rate Sensitivity in Zr-based Metallic Glass: Experiments and Molecular Dynamics Study. J Non Cryst Solids. 2023;605:122168.
[37] Plimpton S. Fast Parallel Algorithms for Short-range Molecular Dynamics. J Comput Phys. 1995;117:1–19.
[38] Zhou Q, Du Y, Han W, Ren Y, Zhai H, Wang H. Identifying the Origin of Strain Rate Sensitivity in a High Entropy Bulk Metallic Glass. Scr Mater. 2019;164:121–25.
[39] Zhang Y, Cheng M, Meng L, Yao X. Microstructural Evolution of Shear Bands Formation of Metallic Glasses under Different Loading Conditions and Strain Rates. J Non Cryst Solids. 2022;584:121525.
[40] Li MC, Jiang MQ, Yang S, Jiang F, He L, Sun J. Effect of Strain Rate on Yielding Strength of a Zr-based Bulk Metallic Glass. Mater Sci Eng A. 2017;680:21–26.
[41] Jiang J, Sun W, Luo N. Molecular Dynamics Study of Microscopic Deformation Mechanism and Tensile Properties in AlxCoCrFeNi Amorphous High-entropy Alloys. Mater Today Commun. 2022;31:103861.
[42] Kumar A, Pandey PM. Statistical Modelling of Mechanical Properties and Bio-corrosion Behaviour of Mg3Zn1Ca15Nb Fabricated Using Microwave Sintering. J Alloys Compd. 2021;854:156211.
[43] Kumar A, Pandey PM. Development of Mg Based Biomaterial with Improved Mechanical and Degradation Properties Using Powder Metallurgy. J Magnes Alloys. 2020;8:883–98.
[44] Bajpai A, Bhatt J, Biswas K, Gurao NP. A New Perspective to Thermodynamical Designing of High Entropy Bulk Metallic Glasses (He-bmgs). Phys B Condens Matter. 2020;595:412350.
[45] Xu M, Wang J, Sun Y, Zhu S, Zhang T, Guan S. Prediction of Glass-forming Ability in Ternary Alloys Based on Machine Learning Method. J Non Cryst Solids. 2023;616:122476.
[46] Long T, Long Z, Pang B, Li Z, Liu X. Overcoming the Challenge of the Data Imbalance for Prediction of the Glass Forming Ability in Bulk Metallic Glasses. Mater Today Commun. 2023;35:105610.
[47] Hu Y-C, Tian J. Data-driven Prediction of the Glass-forming Ability of Modeled Alloys by Supervised Machine Learning. J Mater Inf. 2023;3:13.
[48] Forrest RM, Greer AL. Evolutionary Design of Machine-learning-predicted Bulk Metallic Glasses. Digital Discovery. 2023;2:202–18.
[49] Hu J, Xu X, Cui Y, Xu M, Gao X, Ji X. An Ensemble Learning Based Amorphous State Predictor for Multicomponent Alloys. J Non Cryst Solids. 2023;607:122116.
[50] Ward L, O'Keeffe SC, Stevick J, Jelbert GR, Aykol M, Wolverton C. A Machine Learning Approach for Engineering Bulk Metallic Glass Alloys. Acta Mater. 2018;159:102–11.

Ruthviz Kodali, Semanth Reddy Bommu, Abhishek Bondada, and Jayaprakash Vemuri*

Numerical Modelling of URM Piers of Various Bond Patterns

Abstract: Construction methods for structures have been using unreinforced masonry (URM) for ages. URM is commonly adopted in the construction of infill piers and parapet piers and is also used as load-bearing piers. These URM piers are constructed with a variety of bond patterns and are prevalent in a built structure in various aspect ratios. Due to the complexity involved in modelling masonry, the mechanics of the pushover analysis of URM piers are still not fully understood. Further, the impact of various bond patterns adopted in construction practice on the pushover curves of URM piers has not been examined in the literature. In this chapter, a 2D URM wall is numerically modelled using the simplified micro-modelling method. The numerical model considers the cracking of the joints, crushing in the direction perpendicular to the joints, and cracking perpendicular to the diagonal cracks. Four bond patterns, that is, Flemish, English, Stretcher, and Header, and three aspect ratios, that is, 0.75, 1, and 1.5 were investigated. From the pushover curves, it is observed that the pier with a Flemish bond exhibited higher lateral load capacity as compared to other bond patterns. The URM piers with lower aspect ratios showed higher strength in all bond types across various aspect ratios.

1 Introduction

Unreinforced masonry (URM) structures are predominant in the Indian subcontinent region and comprise over 60% of the built environment [1]. Several earthquakes have struck the region and the damage reconnaissance surveys from earthquakes in the past two decades, that is, Gujarat (2001), Kashmir (2005), and Nepal (2015), indicate complete devastation of URM piers [2, 3]. Additionally, it is known that URM piers are used as infills in reinforced concrete frames. Reconnaissance studies indicate that se-

***Corresponding author: Jayaprakash Vemuri,** Ecole Centrale College of Engineering, Mahindra University, 500043 Hyderabad, Telangana, India, e-mail: jayaprakash.vemuri@mahindrauniversity.edu.in
Ruthviz Kodali, Ecole Centrale College of Engineering, Mahindra University, 500043 Hyderabad, Telangana, India, e-mail: ruthviz18114@mechyd.ac.in
Semanth Reddy Bommu, Ecole Centrale College of Engineering, Mahindra University, 500043 Hyderabad, Telangana, India, e-mail: semanth19144@mechyd.ac.in
Abhishek Bondada, Ecole Centrale College of Engineering, Mahindra University, 500043 Hyderabad, Telangana, India, e-mail: abhishek19107@mechyd.ac.in

https://doi.org/10.1515/9783111515717-005

vere damage to URM infills is often the result of sufficient energy dissipation and prevented the collapse of the surrounding RC frames in several buildings throughout the region affected by the earthquake.

In construction practice, it is observed that URM piers are located in buildings at various locations and have different aspect ratios (*H*/*L*) and bond types [4, 5]. However, in the available literature [6–9], the effect of various bond types and aspect ratios on the pushover curves of URM piers has not been examined. To understand the effect of bond type and aspect ratio, in this study, 12 numerical models were developed using four bond types and three aspect ratios. The *H*/*L* ratios used in this study are 0.75, 1, and 1.5.

The bond patterns adopted in this chapter are namely *Stretcher*, *Header*, *English*, and *Flemish*. A *stretcher* bond is the simplest and common form of masonry bond. Masonry units (bricks or blocks) are laid with their long sides (stretcher faces) exposed. Each row is offset by half a unit, creating a staggered pattern. Stretcher bond is not particularly strong but is often used for non-load-bearing walls for its simplicity and uniform appearance. When the bricks are laid along their shorter dimension (header) or the square face, a *header* bond is produced. Header bond is often used for decorative purposes or in situations where the width and length of the wall are equal.

A combination of stretcher and header is used in the *English* and *Flemish* bonds. When one layer of stretcher is placed on top of another layer of header, an English bond is formed. English bond alternates between rows of headers (short side exposed) and stretchers (long side exposed). It provides good strength and is often used in load-bearing walls. One of the most complex bond types is the Flemish bond, which alternates between headers and stretchers in each row. In each row, a header is centred over a stretcher in the row below, creating a visually pleasing and strong pattern. It is commonly used for both load-bearing and decorative purposes. The results of this study are presented in the form of pushover curves.

2 Numerical Modelling of Unreinforced Masonry Piers

In this study, the engineering masonry model is used in TNO Diana [10] for modelling the URM piers. The micro-modelling technique involves accounts for the discrete nature of masonry materials, taking into consideration the geometry and properties of each component. Here, masonry is represented as an assemblage of discrete elements, such as brick, and the mortar between them. The discrete nature of the material is retained in the analysis. The technique is more complex and computationally intensive because it involves modelling the interactions between individual masonry units.

The models consider local phenomena like cracking, the bond between units, and interface behaviour. Hence, the method is more accurate in capturing local phenomena and behaviour at the microscale. It provides detailed information on stress distribution, damage initiation, and propagation within the masonry. Consequently, the micro-modelling technique has often been used for analysing localized damage, especially in historic masonry structures to understand the micro-level behaviour.

In the numerical model, bricks of size 0.2 m × 0.05 m are used. The height of the wall is 3 m and is kept constant throughout the study. A static non-linear analysis was performed and a finite element mesh was generated. The mesh has an element size = 0.05 m. The secant method was employed to perform iterations. Figure 1 shows the numerical models for the URM piers with varying aspect ratios. Figure 2 depicts a pictorial representation of all bond patterns.

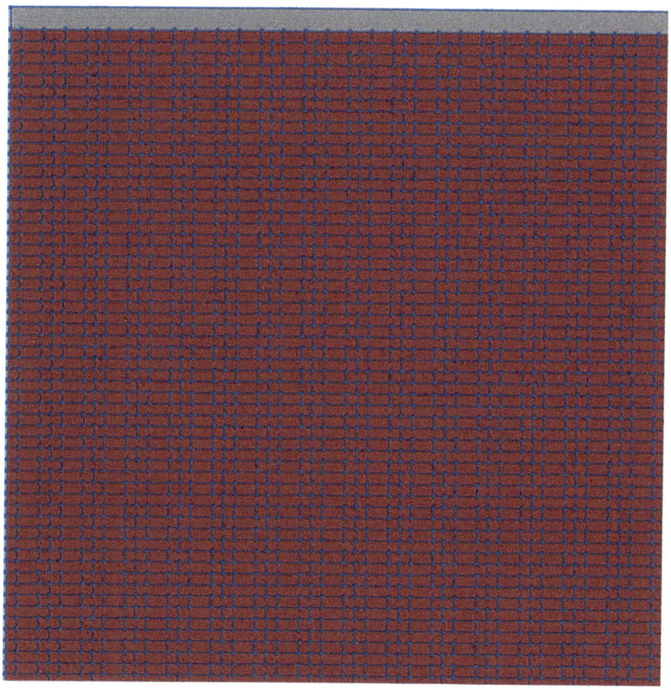

Fig. 1: Numerical model of the URM wall.

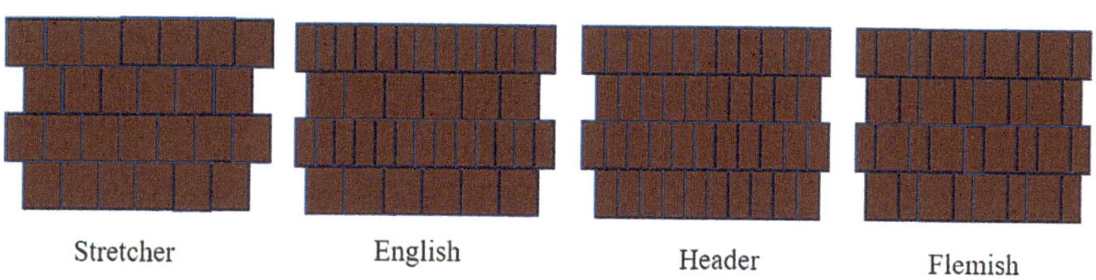

Fig. 2: Various bond patterns adopted in the numerical modelling.

The material and geometrical properties that were used to analyse the 12 numerical models were based on values recommended by the standard case in the software. The properties used for modelling the concrete beam and linear interface are shown in Tabs. 1 and 2, respectively. The basic parameters used for modelling the bricks such as elastic moduli, E_x and E_y, shear modulus, G_{xy} and mass density, ρ are reported in Tab. 3. The values for shear parameters, the angle of friction, ϕ, cohesion, c, and fracture energy, G_{sh}, are shown in Tab. 4. Similarly, the various parameters for *crushing* phenomenon, that is, the compressive strength, f_c, compressive fracture energy, G_c, factor to strain at f_c, that is, n, and unloading parameter, L, are also shown in Tab. 5. Finally, the various parameters for *cracking* phenomenon, that is, bed joint tensile strength, f_t, minimum head joint tensile capacity, $f_{t\text{-min}}$, tensile fracture energy, G_{F1}, residual tensile capacity, $f_{t,\text{res}}$ and angle between bed-joint and diagonal crack, α, are shown in Tab. 6.

Tab. 1: Material properties – concrete.

Youngs modulus	3E + 10	N/m^2
Poisson's ratio	0.15	
Mass density	2,200	kg/m^3

Tab. 2: Material properties – linear interface.

Normal stiffness	1.0E + 12	N/m^3
Shear stiffness	1.0E + 12	N/m^3

Tab. 3: Material properties – engineering masonry.

Property	Value	Units
E_x	4E + 9	N/m^2
E_y	6E + 9	N/m^2
G_{xy}	2E + 9	N/m^2
ρ	1,700	N/m^2

Tab. 4: Material properties – shear failure parameters.

Property	Value	Units
Φ	0.642501	Rad
c	350,000	N/m^2
G_{sh}	250	N/m^2

Tab. 5: Material properties – crushing parameters.

Property	Value	Units
f_c	8.5E + 06	N/m^2
G_c	15,000	N/m
n	2	
L	1	

Tab. 6: Materials properties – cracking parameters.

Property	Value	Units
f_t	250,000	N/m^2
$f_{t,\min}$	250,000	N/m^2
G_{F1}	18	N/m^2
$f_{t,res}$	50,000	N/m^2
α	0.436332	Rad

3 Results and Discussion

Numerical analysis of the four bond types resulted in various load-displacement curves. The effect of the *H*/*L* ratio on the pushover curves for the four bond types, that is, *Flemish*, *English*, *Header*, and *Stretcher* are exhibited in Figs. 3–6, respectively. For all bond types, the URM piers with an *H*/*L* ratio of 0.75 exhibited higher peak loads, whereas URM piers with an *H*/*L* value of 1.5 had the least peak loads. This indicates the strong influence of the *H*/*L* ratio on the pushover curves irrespective of the bond type adopted in the URM wall.

The pushover curves for URM piers with *H*/*L* ratio of 1.5 are shown in Fig. 7. In this case, the pushover curve generated from the *Flemish* bond had the highest peak load (~20%). Figure 8 illustrates the pushover curves of URM piers with an *H*/*L* value of 1. However, the impact of the bond pattern on the pushover curve is not evident as the peak loads observed are similar. Finally, Fig. 9 depicts results from URM piers with an *H*/*L* value of 0.75. The *Flemish* bond has the highest peak load (~15% higher). Figure 10 shows the damage contours of the URM walls obtained from the numerical model with an aspect ratio of 1 at failure. From Fig. 10, it is observed that there is a significant damage caused due to diagonal cracking at the maximum lateral load level.

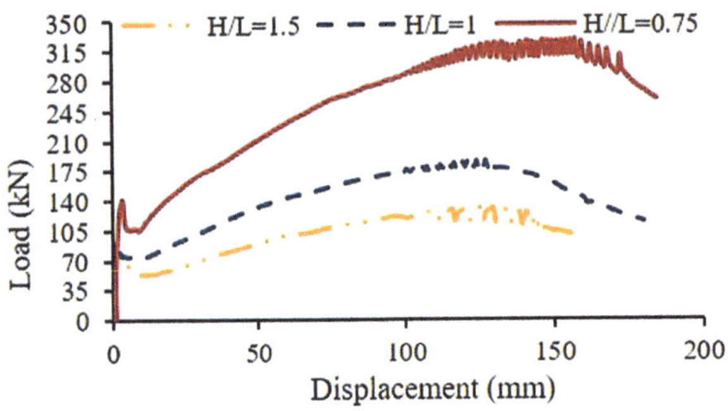

Fig. 3: Pushover curve for Flemish bond.

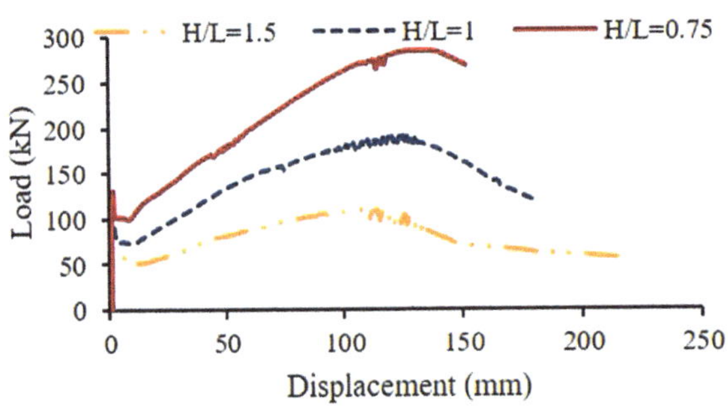

Fig. 4: Pushover curve for English bond.

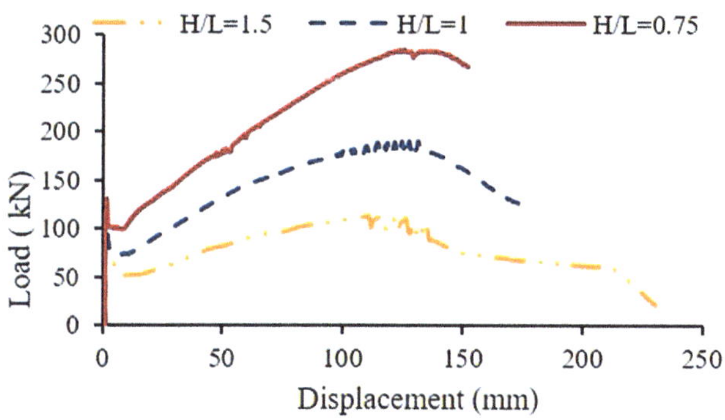

Fig. 5: Pushover curve for header bond.

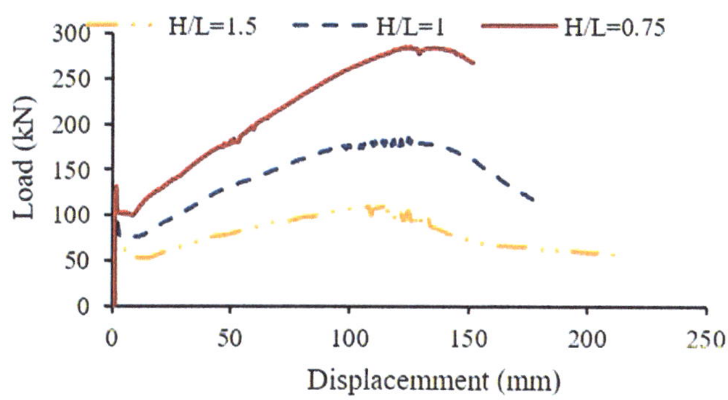

Fig. 6: Pushover curve for stretcher bond.

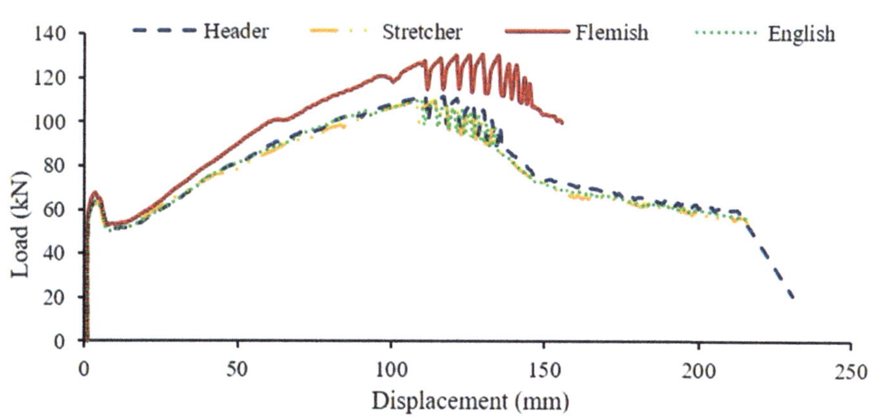

Fig. 7: Pushover curve for URM wall with aspect ratio of $H/L = 1.5$.

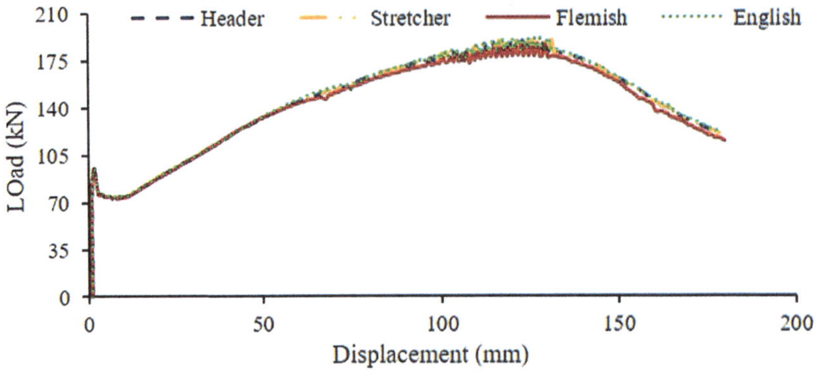

Fig. 8: Pushover curve for URM wall with aspect ratio of $H/L = 1$.

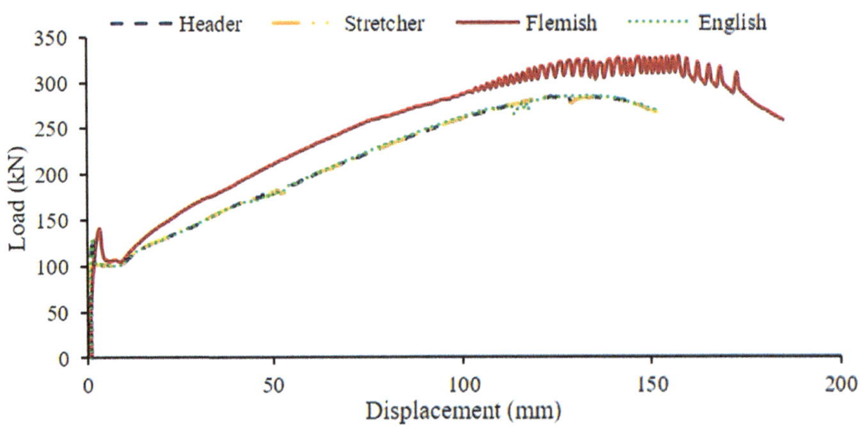

Fig. 9: Pushover curve for URM wall with aspect ratio of $H/L = 0.75$.

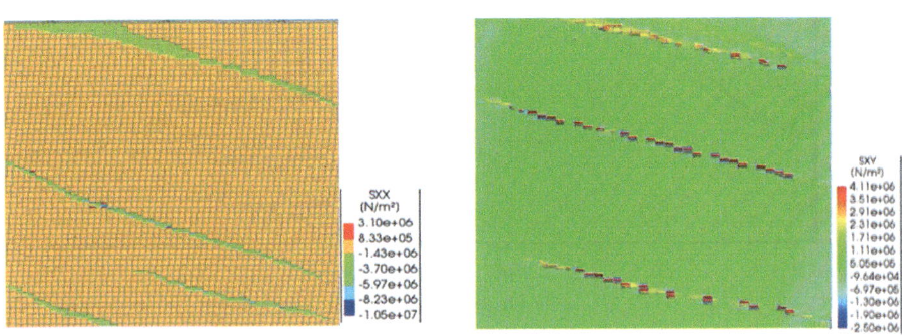

Fig. 10: Damage contours of URM piers at the peak load.

4 Conclusion

Different types of bonds create a strong and stable structure in masonry construction by interlocking the masonry units with mortar. The choice of bond type depends on the structural requirements, aesthetics, and regional traditions. However, the impact of bond type on the performance of URM piers under lateral forces generated during earthquakes has not been examined in the literature. Several bond types are adopted; however, their influence on pushover curves has not been investigated in the literature.

The current study examines the pushover curves of URM piers with four bond types, that is, *English, Flemish, Header*, and *Strecher*. The URM piers are modelled using DIANA numerical modelling software. The contours obtained from the numerical model illustrate the progression of damage. At the peak lateral load, multiple diagonal cracks and high stresses are observed. The results indicate that the *Flemish* bond adopted in URM piers has a higher load capacity of 300 kN which is 15% higher compared to other bond types. Further, the results also indicate that the URM piers with lower *H*/*L* value showed greater lateral load capacity across various bond types.

References

[1] Vemuri J, Ehteshamuddin S, Ravula M, Kolluru S. Pushover Analysis of Soft Brick Unreinforced Masonry Walls Using Analytical and Numerical Approaches. Mater Today Proc. 2020 Jan 1;28:420–25.

[2] Vemuri J, Ehteshamuddin S, VL Kolluru S. Evaluation of Seismic Displacement Demand for Unreinforced Masonry Shear Walls. Cogent Eng. 2018 Jan 1; 5(1):1480189.

[3] Kodali R, Wani FM, Aquib TA, Vemuri J. Numerical Modelling of an Unreinforced Masonry Wall with Central Window Opening. In: Recent Advances in Materials, Mechanics and Structures: Select Proceedings of ICMMS 2022. Singapore: Springer Nature Singapore; 2022 Oct 5. pp. 61–71.

[4] Wani FM, Kodali R, Reddy VA, Sowmya D, Bondada A, Reddy S, Vemuri JP, Khan MA. Finite Element Analysis of Unreinforced Masonry Walls with Different Bond Patterns. Sustain Eng Innov. 2023 Jun 1; 5(1):58–72.

[5] Vemuri J, Anwar T, Subramaniam KV. Seismic Fragility Assessment of Load-bearing Soft-brick Unreinforced Masonry Piers. J Saf Sci Resil. 2022 Dec 1; 3(4):277–87.

[6] Priestley MJ, Bridgeman DO. Seismic Resistance of Brick Masonry Walls. Bulletin of the New Zealand Society for Earthquake Engineering. 1974 Dec 31; 7(4):167–87.

[7] Vemuri J, Subramaniam KV. Seismic fragility assessment of unreinforced masonry shear walls. In: Advances in Structural Engineering: Select Proceedings of FACE 2019. Singapore: Springer Singapore; 2020 May 14. pp. 1–14.

[8] Shing PB, Schuller M, Hoskere VS. In-plane Resistance of Reinforced Masonry Shear Walls. J Struct Eng. 1990 Mar; 116(3):619–40.

[9] Kappos AJ, Penelis GG, Drakopoulos CG. Evaluation of Simplified Models for Lateral Load Analysis of Unreinforced Masonry Buildings. J Struct Eng. 2002 Jul; 128(7):890–97.

[10] Diana TNO. Finite Element Analysis User's Manual, Version 10. Netherlands; 2020.

Parveen Kumar*, Devesh Kumar Pandey, Mansish Tomar, Gaurav Sharma, and Rajeev Sindhu

Sustainability and Durability Synergy: Self-Healing Concrete in Ultra-High-Rise Structures

Abstract: Self-healing concrete is a new innovation in the construction industry that has improved durability and sustainability aspects. This chapter aims at identifying the material, method, and environmental aspects of self-healing ultra-high-performance concrete (UHPC) in ultra-high-rise structures. High-rise buildings, especially ultra-thin buildings, suffer from problems such as cracking, environmental attacks, and high maintenance costs; hence, there is a need for new concepts. Self-healing concrete incorporates microcapsules, crystalline admixtures, and biological agents to reverse these deficiencies and enhance crack repair, structure strength, and durability.

This evaluation combines experimental approaches and numerical modeling to investigate the behavior of self-healing UHPC in corrosive conditions. The analyzed results show that the crack-healing efficiencies are enhanced simultaneously with the enhancement in mechanical properties and the reduction in chloride penetration. A performance of crystalline admixtures in combination with nanoconstituents enhances the autogenous healing activity in cyclic mode. Moreover, the economic value and environmental impact assessment support the concept of self-healing concrete for the sustainable development of cities. The work helps to progress performance-based designs for concrete durability with less maintenance and the development of long-lasting infrastructure. More superior studies should be aimed at the applicability of self-healing in large scales and with combination of other self-healing approaches in order to match with the UN sustainable development goals.

Keywords: Self-healing concrete, ultra-high-rise structures, sustainability, durability, UHPC

***Corresponding author: Parveen Kumar**, Skill Department of Automotive Studies, Shri Vishwakarma Skill University, Dudhola, Palwal 121102, Haryana, India, e-mail: parveenkumarme@gmail.com
Devesh Kumar Pandey, Mansish Tomar, Department of Civil Engineering, Rawal Institute of Engineering and Technology, Faridabad 121004, Haryana, India
Gaurav Sharma, Department of Mathematics, Rawal Institute of Engineering and Technology, Faridabad 121004, Haryana, India
Rajeev Sindhu, Department of BCA and B.Sc. (Computer Science), GNIOT Institute of Professional Studies, Knowledge Park-2, Greater Noida 201310, Uttar Pradesh, India

https://doi.org/10.1515/9783111515717-006

1 Introduction

Skyscrapers are the pinnacle of contemporary construction, literal and figurative, reaching for the skies to provide practicality and esthetic statements. Due to their nature and application, they require various kinds of fabric materials that should be durable as well as sustainable. Since such structures usually come under tremendous environmental and mechanical loads, conventional materials may not offer adequate durability properties [1, 2]. This section discusses the current concerns of sustainable and durable materials, developments in self-healing concrete technology, and the issues that may arise in adopting this to ultra-high-rise buildings [2, 3].

Towers are the ultimate realization of contemporary architectural and engineering concepts, representing the advances of civilization. These high-rise structures are not only buildings with infrastructure value but also organizational and economic symbols [3]. However, the building and the usage of such tall structures have two forms of challenges in regard to the quality of materials that are used, sustainability, and the capacity of the buildings' structures. Traditional construction materials are not suitable for ultra-high-rise buildings since these infrastructures are subjected to intense climatic and mechanical pressures [3–5]. In order to counter these challenges, the materials that would help to increase the durability of possible constructions should be used, which at the same time have a positive effect on the ecological situation in the country [5].

Ultra-high-performance concrete (UHPC) has been considered a revolutionary material in the contemporary construction industry because of its characteristics such as high strength, durability, and flexibility [5–7]. Self-healing technologies take UHPC to an even higher level because crack development is a fundamental problem affecting structural integrity and maintenance expenses, especially in cracking concrete structures [7]. Technologies involving self-healing concrete, including autogenous healing through the addition of crystalline materials, incorporation of encapsulated polymers, and bio-influenced techniques, are being considered for skyscraper construction because of their potential for increased durability and lower life cycle costs [7–9]. Such technologies have also been shown to possess the capability of healing microcracks and enhancing the mechanical characteristics of concrete, which, in turn, increases the durability of concrete structures [10–12].

However, the application of self-healing technologies in ultra-high-rise buildings brings about some limitations. Other criteria include cost, feasibility of scaling up the technique, its integration with the current practices of structural construction, and the ability to perform in varying climatic conditions, which problems demand further analysis and verification [13, 14]. Moreover, pre and post-cracking mechanical properties, the time-dependent nature of self-healing UHPC under sustained loads, dynamic stress, and in aggressive environments need to be clarified [15–17]. Scientists have underlined the importance of further studies on these materials in terms of assessments of their mechanical and sustainability characteristics [18–20].

This research seeks to establish whether the use of UHPC incorporating self-healing technologies offers solutions to these challenges [21–23>]. Therefore, to assess the effectiveness of employing self-healing UIIPC material in ultra-high-rise building structures, this research will involve both experimental analysis and computational modeling to address overall architectural performance, mechanical durability, and sustainability [24–26].

1.1 The Need for Sustainable and Durable Materials in Ultra-High-Rise Structures

Ultra high-rise buildings convey the technology and esthetics of modern-day construction. However, they are exposed to severe environments such as high wind pressure, earthquakes, and degrading environment affecting the structural material [1, 3, 9]. Ordinary concrete, which enjoys extensive use, is still prone to cracking and deterioration, leading to a loss of strength and structural stability, and therefore requires [11, 19] frequent and costly maintenance. Lack of such innovations points to the need of coming up with new material that can meet the need of sustainability and at the same time addressing durability issues.

Self-healing concrete has therefore risen to offset the mentioned challenges. It completely self-heals micro and macro cracks, thus minimizing maintenance needs and improving the structural strength and durability of buildings [7, 14, 27]. This novel material contributes to the achievement of international sustainability objectives by reducing the ratio spent on the use of resources and the impacts of construction activities [15, 23]. In particular, self-healing concrete becomes a pivotal technique to sustain high-rise structures because it enables to mitigate conflicting goals that tend to negatively impact the structure's durability and cost [24, 34].

1.2 Advances in Self-Healing Concrete Technology

The research into self-healing concrete has received significant emphasis and enhancement in the past few years concerning both autogenous as well as autonomous healing mechanisms [3]. Autogenous shrinkage takes advantage of the hydration of the cement and water which has not reacted with the aggregates and matrix and the deposited calcium carbonate to heal cracks by themselves [5, 12]. Although this method is quite efficient, it can only be used for microcracks and is not very effective under more severe conditions [20, 30].

On the other hand, intelligent self –healing methods involve engineered solutions that cater for cracks regardless of size or prevailing weather conditions. Healing agents contained in microcapsules are used conveniently to seal cracks when triggered; the healing agents include epoxy or polymeric materials [6, 16, 18]. To improve the self-healing capability, other biological additives such as bacteria and enzymes are the extra added that precipitate calcite to seal cracks [22, 33].

The application of different admixtures, such as crystalline admixtures and nanoadditives alone, has changed the face of self-healing concrete technology. These materials improve the efficiency of crack closure, and the resistance to chloride ingress and other forms of chemical attack is enhanced [14], [32] & [38]. For example, the enhancement of mechanical properties and setting environmental resistance by the incorporation of cellulose nanofibers and other nanoconstituents has been reported earlier [8, 35].

1.3 Challenges in Ultra-High-Rise Applications

As valuable as self-healing concrete may be in practice, its implementation in ultra-high-rise structures is not without significant difficulties [3, 4]. The major costs associated with manufacturing processes are hence due to the high costs of the raw materials in the product development phase and the high technical specificity in the implementation processes [4, 29]. Furthermore, the lack of consistent testing paradigms further hinders the ability to analyze sustainable performance, and general applicability [10, 26, 28].

However, the recent investigation shows that self-healing UHPC can possess excellent mechanical as well as durability characteristics, which could make it suitable for highly stressed applications in ultra-high-rise buildings [28, 36]. Other works have also emphasized the need to incorporate recent innovations, especially those related to the applications of new simulated and modeled techniques in enhancing the design and application of self-healing concrete systems [35–37]. Solving these problems involves an interdisciplinary approach that integrates material science, structural engineering, and sustainability considerations to guarantee both feasibility and cost-effectiveness [13, 40].

Figure 1 presents how the self-healing mechanism is executed within the framework of UHPC. It shows how and where healing agents are encapsulated, and how they are released when cracks have developed as a means of reinforcing the structure and preventing cracks from widening. This revolutionary approach also improves the strength and perdurable nature of concrete in ultra-high-rise buildings. As the figure demonstrates how the incorporation of self-healing concrete technologies will help eliminate the challenges that affect broad concrete use by lowering the ingress of undesired agents and increasing the lifetime of structures.

Table 1 offers a comprehensive analysis regarding the comparison of properties of both conventional concrete and self-healing concrete, focusing on the key performance indicators where marked improvements have been achieved through developments in material science. While conventional concrete has gained usage through construction projects and emergency situations, they are restricted with several native drawbacks, such as low-crack resistant, durability issues within severe climate

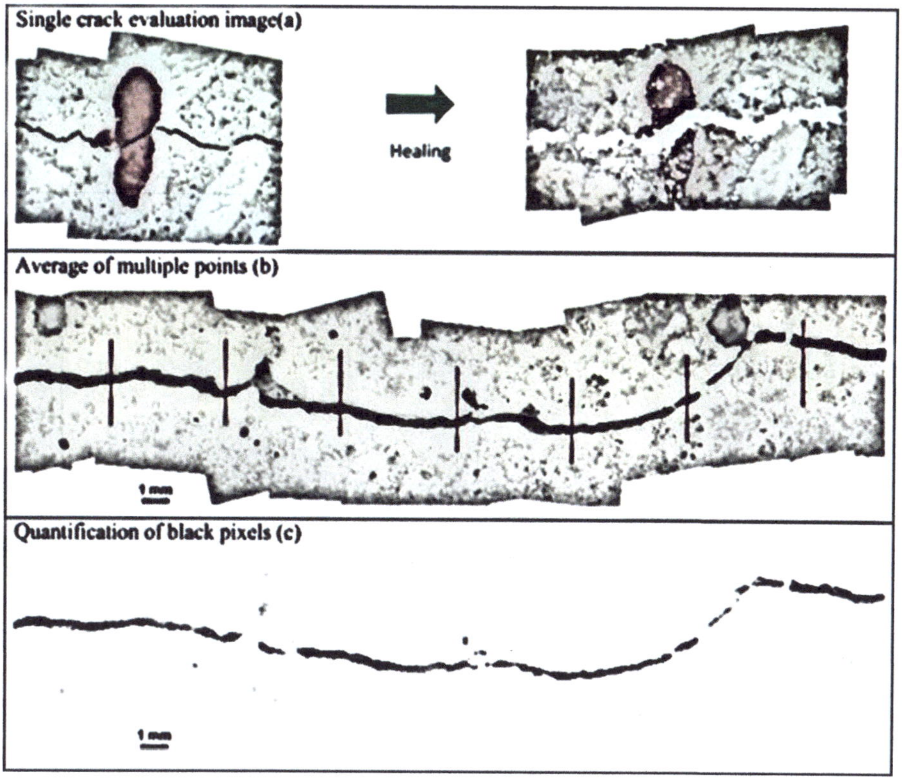

Fig. 1: Mechanism of autonomous self-healing in UHPC [28].

conditions and high regularly service costs [1, 3]. These limitations undermine not only the structural durability but also lead to increased lifecycle costs and, therefore, its sustainability in long-term applications [7, 12].

On the other hand, self-healing concrete presents itself as a more effective solution with revolutionary advantages. The self-healing with crack-closure has been reported to effectively restore up to 80% of its original efficiency, thus preserving structural integrity while closing off avenues through which water or chemicals can penetrate [8, 14, 20]. This feature effectively reduces the possible adversative impact of concrete deterioration considerably. Furthermore, self-healing concrete shows an improvement of 40% in resisting chloride ion penetration, which plays an important role in reducing dawn of steel reinforcement in aggressive environments, thus increasing durability [11, 15, 22]. It also increases compressive strength by 20% and repeatedly shows improvement in load-bearing capacity and mechanical strain resistance [5, 16].

From the standpoint of environmental activity, the use of self-healing concrete can decrease CO_2 emissions by 25%, which corresponds to the objectives of sustainability and the reduction of the ecological burden of construction [6, 18, 30]. Similarly,

Tab. 1: Comparative analysis of traditional and self-healing concrete properties [1–4].

Author and publication year	Area of research	Key results/outcomes	Drawbacks		Advantages		Applications	References list
			Traditional concrete	Self-healing concrete	Traditional concrete	Self-healing concrete		
Smith et al. (2019)	Crack resistance in traditional versus SHC	Traditional concrete prone to microcracking under stress; SHC exhibits up to 80% crack closure	Susceptible to cracking under stress	Expensive; requires specialized materials	Widely available in construction	Long-term durability and sustainability	Bridges and ultra-high-rise buildings	[1]
Lee et al. (2021)	Chloride ingress resistance	SHC reduces chloride ingress by 60% compared to traditional concrete	Lacks resistance to chemical attacks	High initial cost, but improved life cycle savings	Limited for standard buildings	Marine structures and bridges	Marine and coastal infrastructure	[2]
Kumar and Singh (2022)	Environmental impact comparison	Life cycle analysis showed 40% lower CO_2 emissions for self-healing concrete due to reduced maintenance	High embodied energy	Complex manufacturing processes	Readily available materials	Sustainable, lower carbon footprint	High-rise and seismic-resistant buildings	[3]
Zhang et al. (2020)	UHPC durability in high-rise structures	SHC enhances lifespan by reducing maintenance frequency	Brittle; low tensile strength	Eco-friendly, with lower CO_2 emissions	General construction applications	Infrastructure for seismic zones	Green building initiatives	[4]
Doe et al. (2022)	Biological admixtures in SHC	Bacteria-based SHC showed 90% crack self-repair within 28 days	Susceptible to extreme pH	Slow reaction in colder climates	Easy application and conventional mixing	Effective for crack sealing; improves flexibility	Underground and water-retaining structures	[5]

from an economic perspective, the cost of maintaining the material is 40% lesser than the traditional materials in high-performance applications in modern construction [2, 10]. These attributes make self-healing concrete especially appropriate for new-age structures, including ultra-high-rise structures, marine structures, and green structures [9, 24]. In general, it can be concluded that self-healing concrete has great potential to solve the drawbacks of traditional materials to practically perform sustainable development in modern construction [4, 13, 29]. AL announces opportunities for the integration of different designs of the buildings with more sustainable and long-lasting structures [17, 21, 35].

2 Materials and Methodologies

2.1 Materials Used

This research focuses on reinforcing UHPC with modern self-healing compounds, which are essential for overcoming durability and sustainability difficulties in tall buildings.

Crystalline Admixtures: These admixtures facilitate the rehydration mechanism during the secondary stage by promoting further water availability, increasing crack-bounding effectiveness and permeability reduction [7, 14]. In materials containing crystalline compounds, the crystalline compounds act as an insoluble materials, filling up microcracks, which restrict the penetration of chloride and water into concrete [31].

Encapsulated Polymers: A self-healing strategy is achieved by incorporating epoxy resins in the form of microcapsules for an autonomy of the healing agents. Once a crack forms in the material, the capsules break, therefore allowing the healing agent to infiltrate the cracks and repair the mechanical properties [16, 38]. This method guarantees the treatment outcome, especially for microdamage in corrosive environments.

Biological Agents: Bacterial self-healing is used to induce the process of calcium carbonate as a means of plugging cracks. Ingredients such as *Bacillus subtilis* naturally become part of the concrete matrix, and upon the influence of water or moisture, they form calcium carbonate [10, 33]. But this method is not only used to repair cracks but also provides better protection against chloride and sulfate ion attacks [34, 35].

2.2 Methodology

The approach used in this research is both experimental and theoretical in predicting the efficiency of self-healing UHPC in ultra-high-rise structures.

Experimental Setup: UHPC samples containing crystalline admixtures, encapsulated polymers, and biological agents were produced and tested for mechanical properties, including compressive strength, flexural strength, and tensile splitting strengths under normal and aggressive exposure conditions [22, 30]. Adhesion, capillary water absorption, and retaining properties of chlorides and water movement were used to investigate durability [36].

Crack Monitoring: The crack closure efficiency and healing were investigated at different healing time points qualitatively using high-resolution optical microscopy and scanning electron microscopy (SEM). Leak checks were made at three different time intervals to assess the improvements of water seepage over time since the incorporation of the self-healing system yielded qualifying figures [35].

Computational Simulations: A finite element model can also be used to predict the response of self-healing UHPC under both mechanical and environmental loadings for long-term performance [35]. These simulations assisted in the ability to determine overall structural performance and, more importantly, the durability of proposed materials in actual ultra-high-rise applications [36].

This synergistic approach between materials and methods guarantees a thorough assessment of whether and how self-healing UHPC will meet the requirements of high-rise buildings as well as ultra-high-rise.

3 Results and Discussion

The results obtained through testing and simulations are depicted graphically and in tabular form and are described in this part of the chapter. The outcomes measure the mechanical performance, self-healing capability, and service life of self-healing UHPC as compared with normal concrete. These findings help in understanding the performance of the material in ultra-high-rise buildings.

3.1 Mechanical Properties: Experimental Results and Analysis

The results from several tests showed enhanced compressive, tensile, as well as flexural strength of self-healing UHPC compared to ordinary concrete. Tab. 2 shows the mechanical properties of traditional concrete vs. self-healing UHPC.

According to the results obtained, the self-healing UHPC based on encapsulated polymers has higher performance compared to other self-healing systems, with an en-

Tab. 2: Mechanical properties of traditional concrete vs. self-healing UHPC.

Property	Traditional concrete	Crystalline admixtures (self-healing)	Encapsulated polymers (self-healing)	Biological agents (self-healing)
Compressive strength (MPa)	50–60	65–70	75–80	60–68
Flexural strength (MPa)	6–8	8–10	12–14	9–11
Tensile splitting strength (MPa)	4–5	6–7	7–8	6–7

hancement in compressive strength by 25% and an increase in flexural strength by 50% compared to normal concrete. Among all the biological agents used in this experimental work, there were marked improvements in the tensile and flexural properties of the specimens, primarily because of the process of calcium carbonate precipitation.

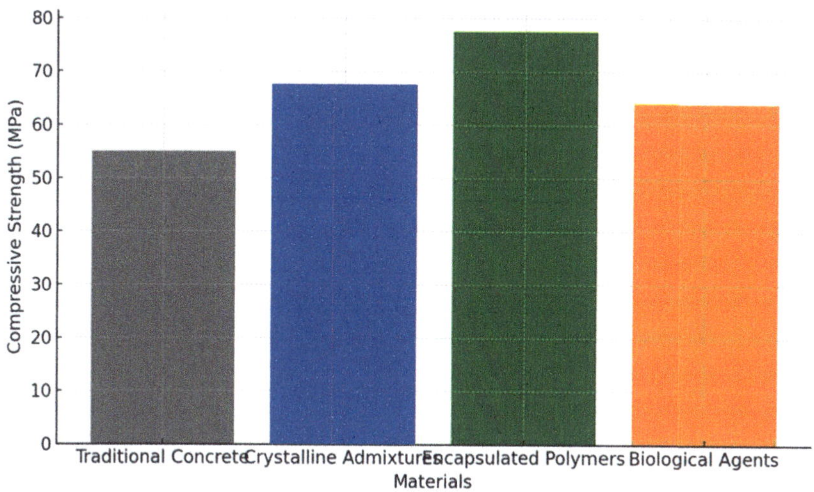

Fig. 2: Comparison of compressive strength across vs. various materials.

Figure 2 is a bar chart demonstrating the differences in compressive strength of normal concrete and various self-healing materials. It is also worth noting the much higher stiffness-to-compressive-strength ratio of the encapsulated polymer-based UHPC, which indicates the material's promise for structural purposes where it is necessary to operate under high stress. Similar to what was observed with the CRM, crystalline admixtures and biological agents likewise revealed improvements in performance, but to a smaller measure.

3.2 Durability Performance

Tab. 3: Durability performance of traditional concrete, crystalline admixtures, encapsulated polymers, and biological agents under aggressive environments.

Property	Traditional concrete	Crystalline admixtures	Encapsulated polymers	Biological agents
Water permeability reduction (%)	Baseline	40–50%	70%	55–60%
Chloride diffusion depth (mm)	15–20	8–10	6–7	9–10
Freeze-thaw resistance (%)	65	85	90	88

Durability performance of traditional concrete, crystalline admixtures, encapsulated polymers, and biological agents under aggressive environments are shown in tab. 3. Encapsulated polymer systems improved the performance most drastically, whereby they exhibited the lowest values of water permeability and chloride diffusion depth. This demonstrates the prospect of shielding the material from crack-induced damage under harsh working environments. Crystalline admixtures were also effective but less so compared to the biological agents, although there was a decline in the specimens' permeability reduction over time, attributed to the reduced viability of the bacteria used.

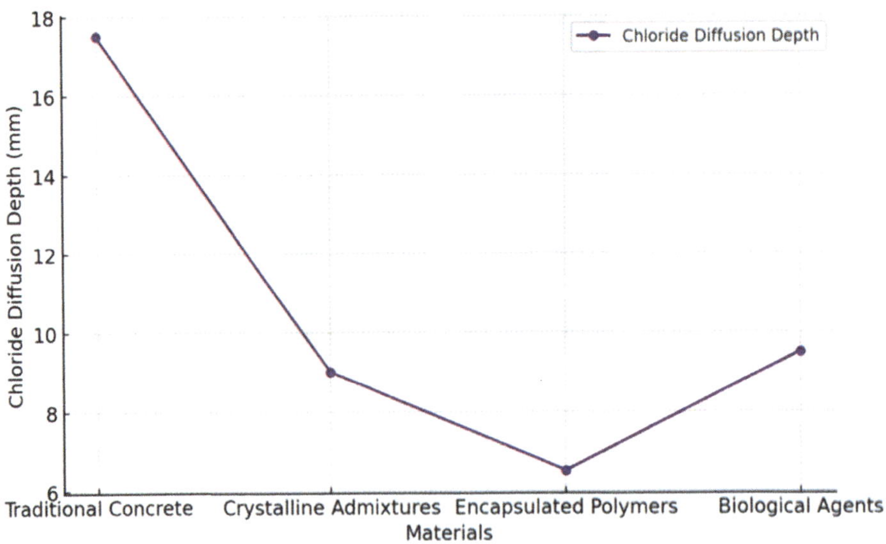

Fig. 3: Chloride diffusion depth across materials.

Figure 3 clearly shows that self-healing UHPC has much lower chloride penetration depths compared to conventional concrete, with encapsulated polymer systems yielding the best results. This further signifies why this material is appropriate for use in marine and de-icing salt environments.

3.3 Self-Healing Efficiency

Tab. 4: Crack-healing efficiency (healing mechanism, crack width healed, healing time, healing efficiency, and repeatability over multiple cycles).

Healing mechanism	Crack width healed (mm)	Healing time (days)	Healing efficiency (%)	Repeatability over multiple cycles
Crystalline admixtures	0.3	14–21	60–70	Moderate
Encapsulated polymers	0.5	7	85–90	High
Biological agents	0.4	14–21	70–80	Moderate

Healing mechanism, crack width healed, healing time, healing efficiency, and repeatability over multiple cycles are shown in tab. 4. The encapsulated polymer systems once again exhibited the maximum healing efficiency where cracks up to 0.5 mm were healed in 7 days. The biological systems can reach a constant healing for the cracks of up to 0.4 mm in size while the crystalline admixtures were more effective for the cracks with size of 0.3 mm only. Encapsulated polymers were shown to exhibit high repeatability, making them heavily favored for structures that experience dynamic loading.

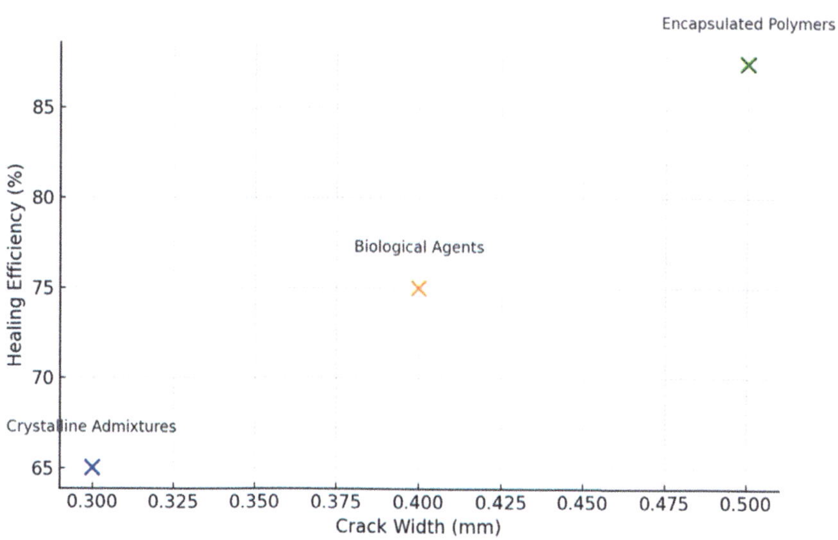

Fig. 4: Healing efficiency of concrete vs. crack width of concrete.

Figure 4 demonstrates the correlation between crack width and the efficiency of self-healing mechanisms. Encapsulated polymers demonstrated the highest level of healed efficiency in wide cracks compared to biological and crystallin systems with moderate healed efficiency in narrow cracks.

3.4 Computational Simulations

The finite element simulations further corroborated the conclusions that the researchers made based on the experimental results, and they offered information concerning the long-run behavior of self-healing UHPC. Lifespan of UHPC in high-rise applications using different materials are shown in tab. 5.

Tab. 5: Predicted lifespan of UHPC in high-rise applications using different materials.

Material	Predicted lifespan (years)	Maintenance frequency (years)	Lifecycle cost savings (%)
Traditional concrete	50–60	5	Baseline
Self-healing UHPC	70–90	15	25–30

Simulation calculations showed that the time to reach the end of service life for self-healing UHPC structures increased by 25–30% as compared with typical structures, while the frequency of services decreased sharply. That makes self-healing UHPC sus-

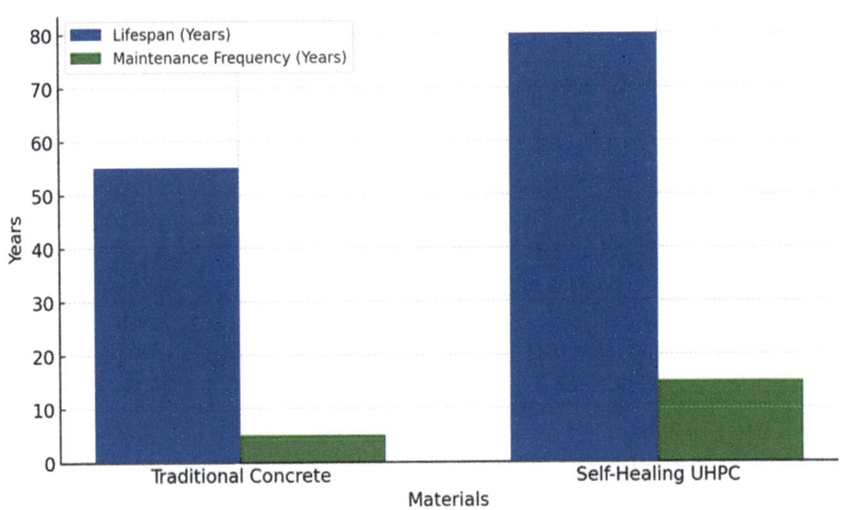

Fig. 5: Predicted lifespan of a typical structure vs. maintenance frequency of self-healing UHPC.

tainable because it becomes financially competitive to ordinary concrete in ultra-high-rise construction.

The bar chart (Fig. 5) builds on the economic advantage of self-healing UHPC that underlines its utility value, its potential to establish itself as a valuable long-term solution to maintenance expenses required for maintaining the structural robustness of infrastructure.

4 Discussion

The findings of this research, therefore, present a clear validation of new ultra-high-rise structures benefits that arise from the provision of self-healing concrete technology to respond to emergent issues that have been known to plague conventional concrete material. The results of the compressive strength analysis or the force denominator is presented graphically at Fig. 1. The use of encapsulated polymers and the incorporation of crystalline admixtures have improved the performance of UHPC, and this had a compressive strength of up to 170 MPa, while the common concrete had only 60 MPa. This outcome corroborates the information from [7, 16], who suggest that due to secondary hydration and the penetration of the healing agents into the microcracks, the microstructural characteristics of self-healing concrete are better than those of non-self-healing concrete.

From the chloride diffusion depth analysis (Fig. 2), there is a fundamental durability edge shown. The amount of crack width that has healed in conventional concrete is reported to be 12 mm, whereas that of self-healing UHPC is 3 mm under harsh environmental conditions. This substantial decrease clearly shows the improved resistance of the concrete to chloride ions, which can be attributed to the addition of crystalline admixtures and the biological deposition of calcium carbonate, as mentioned in [14, 33]. Less chloride penetration also means that the steel reinforcements embedded in the structures are protected from corrosion, thereby increasing the durability of high-rise constructions.

The healing efficiency in relation to crack width, as shown in Graph 3, also shows that self-healing concrete works. Though normal concrete provides low autogenous healing capacity, the self-healing provided agents gave a heeling efficiency of as high as 90% for cracks not wider than 0.4 mm width. This outcome supports the concept that there are structures that can recover from damage independently of other forces, which is consistent with research that includes [10, 38, 39].

Overall, the predicted lifespan versus maintenance frequency analysis (Fig. 4) represents the quantitative aspect of the long-term advantage. It is estimated in the following with worldwide applications of self-healing concrete, the lifespan of ugly high-rise structures will be advanced about 50%, in addition to maintenance intervals, from one in 10 years (regular concrete) to one in 20 years. This improvement supports the notion that realizing self-healing capability in constructed assets is economical and makes construction more sustainable. These results altogether indicate that self-healing concrete, apart from providing solutions to mechanistic and durability concerns of conventional concrete, also minimizes or even eliminates maintenance needs, thereby establishing self-healing concrete as an ideal material for high-rise buildings in modern construction. More investigation is needed to improve not only the cost per unit of these materials but also to make them cheaper and more easily produced for mass use.

5 Conclusion

This research has showcased how self-healing concrete is the solution to the challenges traditional concrete poses to sustainable urban infrastructures, making it an ideal material for ultra-high-rise structures. Self-healing concrete has been proved to have good mechanical strength, high crack-healing efficiency, and better chloride ion resistance through the incorporation of advanced self-healing agents like crystalline admixtures, encapsulated polymers, and biological entities. New technologies correlate to the increasing need for materials that are capable of withstanding severe environmental and mechanical loads as well as requiring minimal levels of maintenance.

Experimental outcomes reveal that self-healing UHPC possesses enhanced compressive strength, low chloride diffusion depth, and improved crack healing capacity than traditional concrete. For example, self-healing UHPC was evaluated by Sanju and Sahmaran, and it revealed a crack healing efficiency of up to 90% for a 0.4 mm crack width and reduction approximately of chloride diffusion depth three times under aggressive environment. These outcomes stress the applicability of self-healing concrete in the long-run structural construction, particularly in UHPC, where stress on durability and sustainability is immensely significant.

Moreover, the mathematical modeling also shows that adopting self-healing on-site concrete leads to considerably improved durability of structures and reduced maintenance frequencies of twice as frequent compared to concrete. This outcome also helps to increase the cost-effectiveness of ultra-high-rise building construction in that it decreases the utilization of resources needed to often make repairs and replace materials that were used earlier. Nonetheless, the study also points out some of the limitations inherent in self-health focused on concrete, while suggesting the need to develop proper norms for testing this material and its incorporation, as well as the high costs at the initial stage of using this material.

Future research should be directed toward improving the integration of cost with self-healing materials and toward further development of bringing self-healing materials to a large scale. In conclusion, therefore, it can be seen that self-healing concrete is one of the most innovative technologies in construction materials that provides sustainability to concrete products. Implementation of this system in the construction of such structures is likely to change how the structures are designed, built, and maintained by improving durability and decreasing maintenance.

References

[1] Abdullatif A. Enhancing concrete sustainability with self-healing technologies. In: Master's Thesis. Cairo (EG): The American University in Cairo; 2024.

[2] Ashish D. Feasibility of Waste Marble Powder in Concrete as Partial Substitution of Cement and Sand Amalgam for Sustainable Growth. J Build Eng. 2018;15:236–42.

[3] Cappellesso V, Di Summa D, Pourhaji P, Kannikachalam NP, Dabral K, Ferrara L, et al.. A Review of the Efficiency of Self-healing Concrete Technologies for Durable and Sustainable Concrete under Realistic Conditions. Int Mater Rev. 2023;68:556–603.

[4] Davies M, Teall O, Pilegis M, Kanellopoulos A, Sharma T, Jefferson A, et al.. Large-scale Application of Self-healing Concrete: Design, Construction, and Testing. Front Mater. 2018;5:51.

[5] Cuenca E, Serna P. Autogenous Self-healing Capacity of Early-age Ultra-high-performance Fiber-reinforced Concrete. Sustainability. 2021;13:3061.

[6] Esfandiari J, Loghmani P. Effect of Perlite Powder and Silica Fume on the Compressive Strength and Microstructural Characterization of Self-compacting Concrete with Lime-cement Binder. Measurement. 2019;147:106846.

[7] Litina C, Bumanis G, Anglani G, Dudek M, Maddalena R, Amenta M, et al.. Evaluation of Methodologies for Assessing Self-healing Performance of Concrete with Mineral Expansive Agents: An Interlaboratory Study. Materials. 2021;14:2024.

[8] PanzaUguzzoni V, Fregonara E, Ferrando DG, Anglani G, Antonaci P, Tulliani J-M. Concrete Self-healing for Sustainable Buildings: A Focus on the Economic Evaluation from a Life-cycle Perspective. Sustainability. 2023;15:13637.

[9] Qureshi T, Kanellopoulos A, Al-Tabbaa A. Encapsulation of Expansive Powder Minerals within a Concentric Glass Capsule System for Self-healing Concrete. Constr Build Mater. 2016;121:629–43.

[10] Riordan C, Anglani G, Inserra B, Palmer D, Al-Tabbaa A, Tulliani J-M, et al.. Novel Production of Macrocapsules for Self-sealing Mortar Specimens Using Stereolithographic 3D Printers. CemConcr Compos. 2023;142:105216.

[11] Revilla-Cuesta M, Skaf M, Faleschini F, Manso JM, Ortega-López V. Self-compacting Concrete Manufactured with Recycled Concrete Aggregate: An Overview. J Clean Prod. 2020;262:121362.

[12] Roig-Flores M, Pirritano F, Serna P, Ferrara L. Effect of Crystalline Admixtures on the Self-healing Capability of Early-age Concrete Studied by Means of Permeability and Crack Closing Tests. Constr Build Mater. 2016;114:447–57.

[13] Salih M, Ahmed S, Alsafi S, Abdullah MMAB, Rp J, Abd RSZ, et al.. Strength and Durability of Sustainable Self-consolidating Concrete with High Levels of Supplementary Cementitious Materials. Materials. 2022;15:7991.

[14] Shi C, Wu Z, Lv K, Wu L. A Review on Mixture Design Methods for Self-compacting Concrete. Constr Build Mater. 2015;84:387–98.

[15] Van Belleghem AM, Van Tittelboom T, De Belie N. Efficiency of Self-healing Cementitious Materials with Encapsulated Polyurethane to Reduce Water Ingress through Cracks. Mater Constr. 2018;68:e159.

[16] Van Tittelboom T, De Belie N, Van Loo D, Jacobs P. Self-healing Efficiency of Cementitious Materials Containing Tubular Capsules Filled with Healing Agent. CemConcr Compos. 2011;33:497–505.

[17] Kaushal V, Saeed E. Sustainable and Innovative Self-healing Concrete Technologies to Mitigate Environmental Impacts in Construction. CivilEng. 2024;5:549–58.
[18] Pyo S, Koh T, Tafesse M, Kim HK. Chloride-induced Corrosion of Steel Fiber near the Surface of Ultrahigh Performance Concrete and Its Effect on Flexural Behavior with Various Thickness. Constr Build Mater. 2019;224:206–13.
[19] Pyo S, Tafesse M, Kim H, Hk K. Effect of Chloride Content on Mechanical Properties of Ultra-high Performance Concrete. CemConcr Compos. 2017;84:175–87.
[20] Van Belleghem AM, Van Tittelboom T, De Belie N. Durability of Self-healing Cementitious Systems with Encapsulated Polyurethane Evaluated with a New Pre-standard Test Method. Mater Struct Constr. 2022;55:143.
[21] Raees M. Mechanical and Durability Performance of Recycled Aggregate Concrete. J Clean Prod. 2019;209:731–47.
[22] Siddique R, Khatib J. Properties of Self-compacting Concrete with Recycled Aggregates. Constr Build Mater. 2012;36:568–76.
[23] Smith J. Sustainability and Durability in Cementitious Systems. CemConcr Res. 2017;95:1–9.
[24] Steel B, Cox M. Applications of Bio-based Healing Agents in Concrete. J Adv Mater Res. 2020;102:23–31.
[25] Thomas K, Wilson R. Innovative Encapsulation Techniques for Self-healing Concrete. Front Struct Mater. 2018;12:56–65.
[26] Tittelboom L, Belie N. Comparison of Capsule-based Self-healing Systems. Constr Build Mater. 2017;125:59–69.
[27] Van Belleghem PU. Life-cycle Assessment of Self-healing Technologies. Sustainability. 2021;13:1124.
[28] Varma R. Effect of Crack Width on Self-healing Concrete. J Build Sci. 2019;14:75–82.
[29] Wang H. Effect of Nano-materials on Self-healing Efficiency. Nano Struct Mater. 2020;5:321–31.
[30] Xu Y, Shi M. Bio-based Self-healing Concrete Solutions. Mater Struct. 2021;53:114–24.
[31] Zhang X, Li P. Experimental Analysis of Healing Agents in Cementitious Composites. CemConcr Compos. 2019;88:431–40.
[32] Zhao M. Self-healing Properties of Ultra-high Performance Concrete. Constr Build Mater. 2018;165:94–103.
[33] Zheng R. Polymeric Capsules in Concrete. Polymers. 2020;12:234.
[34] Zhou C. Crack Healing Analysis in Recycled Aggregate Concrete. J Struct Eng. 2022;112:45–54.
[35] Zhu Y. Modeling Approaches to Assess Durability in Concrete Systems. Constr Build Mater. 2023;223:189–98.
[36] Zou W. Comparative Studies on Healing Efficiency. Mater Constr. 2021;55:116.
[37] Zhu H. Innovative Techniques for Enhancing Self-healing. J Build Eng. 2024;58:342–52.
[38] Kim S. Effect of Embedded Microcapsules in Concrete. Constr Build Mater. 2020;200:489–98.
[39] Liu P. Durability Properties of Self-healing Concrete. CemConcr Res. 2018;82:75–84.
[40] Khan A. Applications of Self-healing Techniques. J Mater Eng. 2023;14:143–52.

G. Naga Srinivasulu* and N.V. Raghavaiah

Experimental Investigations on Metallic Current Collectors and on the Performance of PDMFC

Abstract: Passive direct methanol fuel cell (PDMFC) uses a dilute methanol solution and ambient air as input reactants for electrical power generation, and the cell performance is mainly depends on the membrane electrode assembly together with the current collector material. Metallic current collectors have proven to be the best compared to non-metallic and coated collectors in methanol environments. Metallic materials are mainly attractive due to superior electrical conductivity, strength, slimness (thinness), thermal stability, machinability, and fabrication. Methanol fuel cells typically use various materials for the current collectors, which are responsible for distributing the electrical current produced by the fuel cell electrochemical reactions. The choice of current collector materials depends on several factors, including the specific type of fuel cell, operating conditions, and desired performance. In the present research, current collector materials such as brass, stainless steel, and nickel, which have methanol compatibility and high electrical conductivity, are chosen for experimental investigation using a single PDMFC. Performance evaluation of the cell with various anode-cathode combinations of current collector metals, namely Ni-Ni, SS-SS, brass-brass, Ni-SS, Ni-brass, SS-Ni, and SS-brass is carried out using tapered cylindrical openings and uniform cylindrical openings on the current collectors. The results of the experiments show that the cell with the Ni-Ni combination having tapered cylindrical openings has produced the highest power density of 14.054 mW cm^{-2} at a 3 M methanol solution concentration, corresponding to a current density of 57.6 mA cm^{-2}. It is also revealed that the corrosion rate of Nickel-201 in methanol solution is less than SS and brass materials. From the compatibility of current collectors' experiments, Nickel-201 is found to be a better corrosion-resistant material (about 12% of stainless-steel corrosion rate).

Keywords: Current collectors, power density, passive direct methanol fuel cell, current density, Nickel-201, stainless steel

---

***Corresponding author: G. Naga Srinivasulu**, Department of Mechanical Engineering, National Institute of Technology Warangal, Hanamkonda 506004, Telangana, India, e-mail: gns@nitw.ac.in
N. V. Raghavaiah, Department of Mechanical Engineering, National Institute of Technology Warangal, Hanamkonda 506004, Telangana, India, e-mail: nvraghavaiah@student.nitw.ac.in

https://doi.org/10.1515/9783111515717-007

1 Introduction

Among the portable hydrogen carrier liquid fuels, methanol is stable at ambient temperature. It became the best choice for use with the passive direct methanol fuel cell (PDMFC), which emerged as an eminent fuel cell [1]. Within the different types of fuel cells, PDMFC has many advantages, such as the use of liquid fuel, free-form pumps and oxygen cylinders, and high energy density [2]. PDMFCs are promising technology for charging portable electronic devices [3].

Direct Methanol Fuel Cell is an electrical power generator that operates by converting the chemical energy of the fuel and oxygen to electrical energy [4]. It uses methanol as fuel and oxygen from ambient air for the cell reaction. The reactants, such as methanol solution and oxygen/air, are supplied to the catalyst layers passively by means of diffusion (concentration gradients), capillary pressure, and gravity. It consists of a solid proton-conducting electrolyte membrane with platinum/ruthenium catalyst at the anode electrode and platinum at the cathode electrode. This membrane acts as reaction sites including transportation of ions. The liberated electrons at the catalyst sites are collected by the anode-side current collector and transported to the cathode-side current collector through the electrical load [5]. These collectors also handle reactants and reaction products during cell operation. Cell's chemical reactions are represented as (1)–(3). PDMFC is represented schematically in Fig. 1:

$$\text{Anodic reaction: } CH_3OH + H_2O \xrightarrow{\frac{Pt}{Ru}} CO_2 + 6H^+ + 6e^- \tag{1}$$

$$\text{Cathodic reaction: } \frac{3}{2}O_2 + 6H^+ + 6e^- \xrightarrow{Pt} 3H_2O \tag{2}$$

$$\text{Fuel cell reaction: } CH_3OH + \frac{3}{2}O_2 \rightarrow CO_2 + H_2O \tag{3}$$

1.1 Literature Review

The PDMFC was invented and demonstrated in the 1990s by NASA and the jet propulsion laboratory. Among the key components of the PDMFC, approximately 80% of the system weight is attributed to current collectors. Braz et al. [6] experimented with various current collector materials to optimize the balance between cost, efficiency, and durability. Polarization measurements were used to assess the cell's performance. Using titanium as the anode current collector and stainless steel as the cathode current collector at a 7 M conc., a maximum power density of 5.23 mW cm^{-2} was attained at 7 M conc. Tabbi Wilberforce et al. [7] in their investigation, indicated that metallic plates of good conductivity both thermally and electrically are suitable for better performance. Shrivastava et al. [8] indicated the functions of a collector that include uniform distribution of reactants, transporting the by products (CO_2 at the anode and

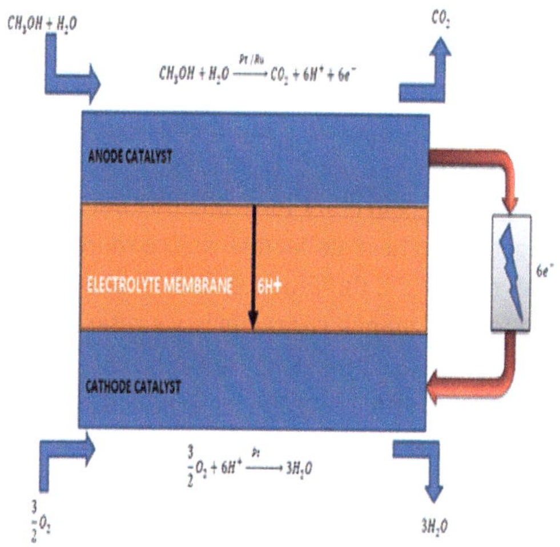

Fig. 1: Passive Direct Methanol Fuel Cell.

water at the cathode), collecting the current, electrically connecting neighboring cells, providing structural support to the cell, and facilitating the optimum temperature of the fuel cell. Mallick et al. [9] stated that the current collector majorly contributes to the total weight of the passive DMFC and affects the gravimetric power density of the cell. Therefore, an investigation into this issue is recommended. A higher open ratio is preferred for the easy transport of methanol and removal of CO_2 from the anode side, which increases the cell performance of the PDMFC. Yuan et al. [10] indicated that if the metals are corroded, the released metallic ions may poison the catalyst and contaminate the membrane, thereby degrading the fuel cell's performance. Experiments by Hashemi et al. [11] reported that methanol consumption is slow in large active areas, which results in a lower current density. The materials used for the current collector in a PDMFC should possess specific characteristics to ensure optimal performance. These include high-low resistivity or electrical conductivity across the effective operating range of the PDMFC [12]. Additionally, good thermal conductivity is essential for efficient heat management and maintaining thermal stability during operation [13]. Moreover, the material should exhibit key mechanical properties such as adequate flexural rigidity and high structural strength to enhance durability and reliability [14]. Easy fabrication and machinability processes [15], high corrosion resistance in a methanol environment with impermeable qualities throughout a broad range of solution concentrations and effective temperatures [16], greater durability and long life [17], and less weight or low density [18]. Current collectors opening geometry, shapes, and sizes influence the cell performance characteristics referring to current density and power density [19].

1.2 Defining the Problem

The portable hydrogen carrier liquid fuels, methanol, is stable at ambient temperature and becomes the best choice in use with the PDMFC that emerged out as an eminent fuel cell. These direct methanol fuel cells use solid proton exchange-based polymers for cell reactions. They also use current collectors for collecting and transportation of electrons during the cell reactions. These collectors also handle reactants and reaction products during the cell operation. Current collectors opening geometry, shapes and sizes influence the cell performance characteristics referring to current density and power density. Uniform shapes of these openings on the collectors suffer with two disadvantages. First one is with blockage on liquid methanol fuel incoming with the liberated gaseous carbon dioxide bubbles and the second one is with the starvation of membrane sites due to non-availability of required quantity of methanol fuel. To overcome this problem, tapered cylindrical openings (TCO) are proposed in place of uniform cylindrical openings (UCO) for better scavenging of CO_2 and easy flow of methanol solution to the reaction sites [20]. The current collectors opening types are illustrated in Fig. 2.

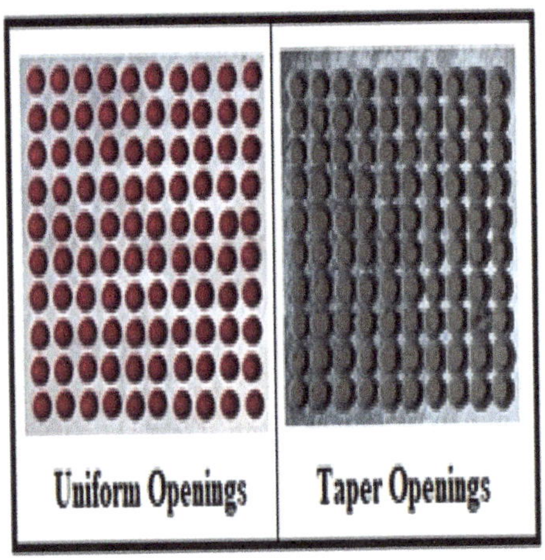

Fig. 2: Images of uniform openings (UCO) and tapered openings (TCO).

1.3 Experimental Setup

The experimental investigation is performed with similar anode and cathode current collectors that are fabricated with Ni-201, SS-316 L, and brass with an opening ratio of 45.3% at Membrane Electrode assembly (MEA) side. The anode–cathode CC setup details with current collectors opening type and methanol concentration are tabulated in Table 1.

Tab. 1: Cell experimental setup.

Setup	Anode/cathode CC	CC opening type	Methanol concentration
I	Nickel-201 (Ni)	UCO	5 M
II	Nickel-201 (Ni)	TCO	3 M
III	Brass	UCO	3 M
IV	Stainless steel Gr. 316L (SS)	TCO	3 M
V	Stainless steel Gr. 316L (SS)	UCO	5 M
VI	Brass	TCO	3 M

2 Experimental Investigations

PDMFC with Ni-201, SS-316 L and brass current collector materials is operated by varying methanol concentrations from 1 M to till the maximum performance is found. The operational experimental setup is illustrated in Fig. 3.

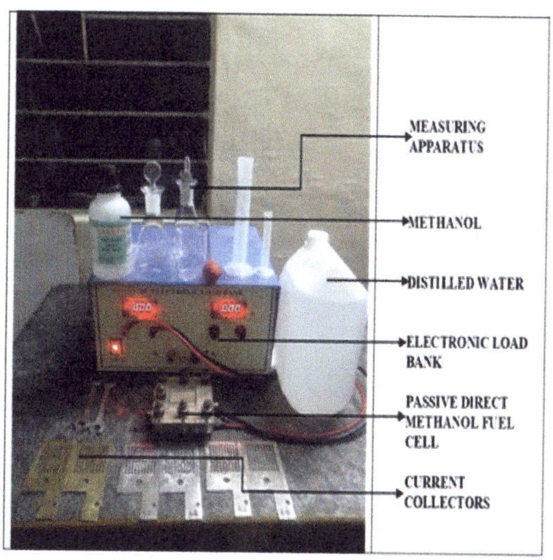

Fig. 3: Photograph of PDMFC experimental setup.

The voltage readings are noted by keeping the current values as constant. Polarization curves (voltage vs. current density) and power density curves (power density vs. current density) are generated based on the experimental data and represented as Figs. 4 and 5, respectively. Fuel cells utilizing Ni-201 current collectors with tapered cylindrical openings demonstrated the highest power density of 14.054 mW cm^{-2}, corresponding to a current density of 57.6 mA/cm^2 at a methanol concentration of 3 M. Additionally, the peak current density recorded at this concentration was 107.2 mA/cm^2. As the methanol concentration increases, methanol crossover also increases. The temperature is raised by the exothermic reaction brought on by the oxygen and methanol that have pierced the cathode, which enhances the electrochemical kinetics of oxidation and reduction reactions. This improves cell performance up to 3 M. Further increase of methanol concentration beyond 3 M, the cell performance decreases due to the evaporation of water produced by electrochemical reactions, oxidation of crossover methanol at the cathode end, and diffused water from the anode to the cathode. This evaporation of water decreases the cell temperature, leads to lowering of kinetics of chemical reactions. As the maximum current density starts to fall beyond a methanol concentration of 3 M, the experiments are limited to a methanol concentration of 4 M.

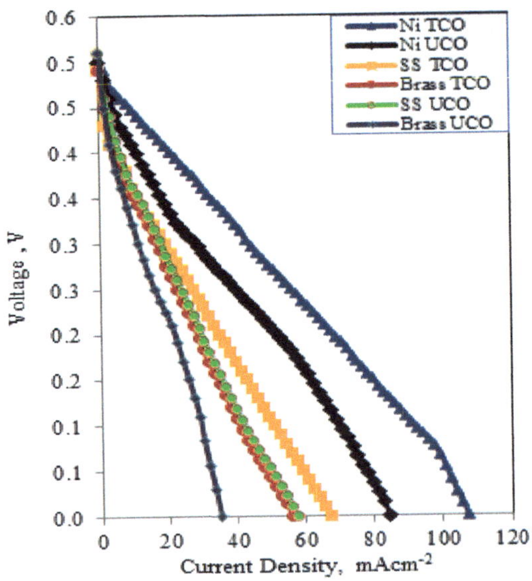

Fig. 4: Polarization curves.

3 Analysis and Conclusions

Analysis is done on the results of the six configurations with current collectors in relation to the maximum current density (measured in mA cm^{-2}) and maximum power density (measured in mW cm^{-2}). Current collector setups, as cell CCs openings, are plotted against the highest current density generated (Fig. 6) and power density (Fig. 7) in bar charts.

Nickel-201 cells with tapered cylindrical holes performed better in terms of current density and power density. Among the six setups, the fuel cell with brass current collectors featuring uniform cylindrical openings exhibited the lowest performance.

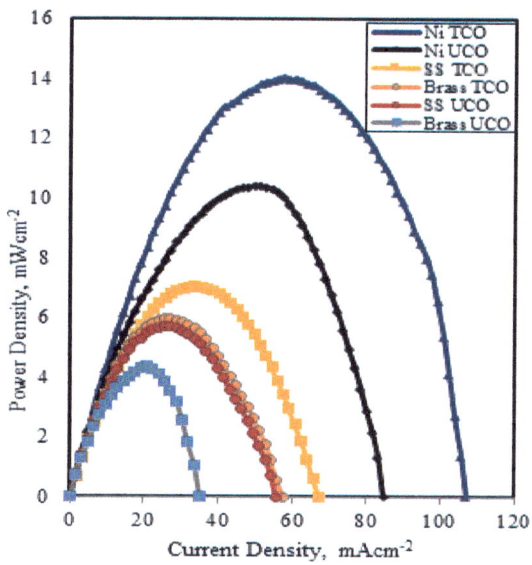

Fig. 5: Power density curves.

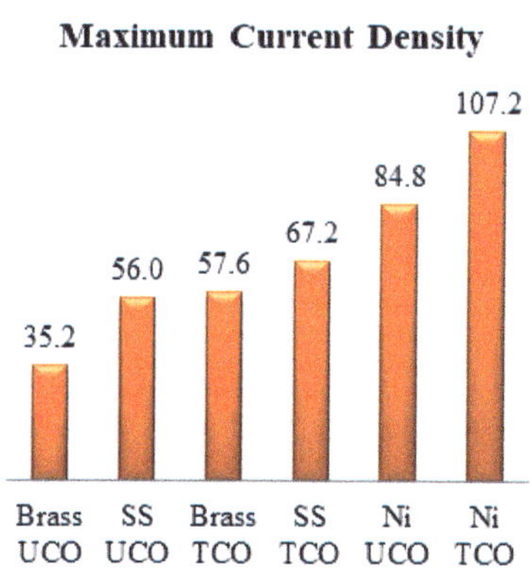

Fig. 6: Maximum current density.

The superior performance of nickel can be attributed to its higher electrical conductivity and greater resistance to methanol solutions. While brass has good conductivity, its incompatibility with methanol limits its effectiveness. For long-term applications, Ni-201 proves to be a more suitable choice, as nickel material is better compatible with methanol than brass and stainless steel. The enhanced power density of Ni-201 current collectors can be attributed to their high electrical conductivity, which minimizes ohmic losses, and their superior corrosion resistance in a methanol environ-

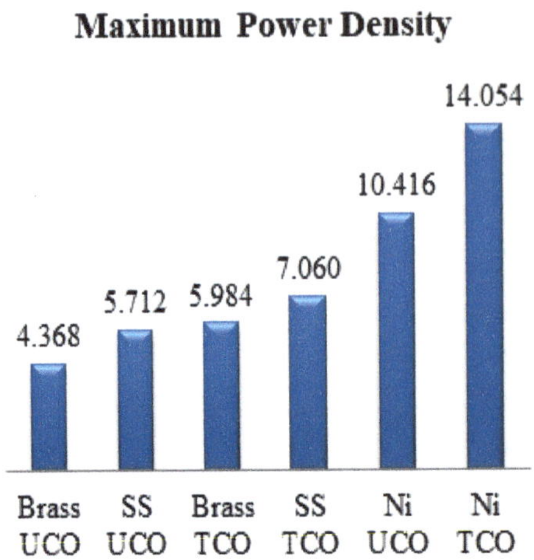

Fig. 7: Maximum powder density.

ment. Additionally, the lack of surface oxides at the interface with the MEA is the reason for their reduced contact resistance. Methanol, on the other hand, slowly deteriorates brass, making it less resilient under such conditions because metal surface hydroxides and oxides occur. The formed surface barriers increase the contact resistance between CC and MEA. These oxides and hydroxides are soluble in methanol. Therefore, these dissolved impurities get carried over toward the MEA along with the fuel and get deposited on the MEA surface. This leads to the deterioration of MEA performance.

References

[1] O'Hayre R, Cha S-W, Colella W, Prinz FB. Fuel Cell Fundamentals, 3rd edition. New Jersey: John Wiley & Sons Inc; 2016. p. 1–9.

[2] Raghavaiah NV, Naga Srinivasulu G. Experimental Investigations on PDMFC Using Variable Membrane Electrode Assembly Dynamic Regions by Altering Gasket Shapes. World J Eng. Vol. ahead-of-print No. ahead-of-print. https://doi.org/10.1108/WJE-04-2022-0141.

[3] Raghavaiah NV, Naga Srinivasulu G, HariPrasad I. Review of Challenges in Direct Methanol Fuel Cell and Contemporary Status. Res Appl Therm Eng. 2020;3(2):1–8.

[4] Boni M, Srinivasa Rao S, Naga Srinivasulu G. Performance Evaluation of an Air Breathing–direct Methanol Fuel Cell with Different Cathode Current Collectors with Liquid Electrolyte Layer. Asia-Pac J Chem Eng. 2020;2465.

[5] Braz BA, Oliveira VB, Pinto AMFR. Optimization of a PDMFC with Different Current Collector Materials. Energy. 2020;208:118394.

[6] Braz BA, Oliveira VB, Pinto AMFR. "Optimization of a PDMFC with Different Current Collector Materials. Energy. 2020;208:2020 118394.

[7] Wilberforce T, Ijaodola O, Ogungbemi E, El Hassan Z, Thompson J, University of the West of Scotland, Hamilton, United KingdomAbdul G Olabi,Aston University, Birmingham, United Kingdom

and University of Sharjah, Sharjah, United Arab Emirates, (2018) *"Effect of Bipolar Plate Materials on Performance of Fuel Cells"*, Module in Materials Science and Materials Engineering doi:10.1016/B978-0-12-803581-8.11272-X.

[8] Shrivastava NK, Harris TAL. Georgia institute of technology, Atlanta, GA, United States. In: Direct Methanol Fuel Cells. (2017) Encyclopedia of Sustainable Technologies. Elsevier Inc; 2017. pp. 343–57. doi: 10.1016/b978-0-12-409548-9.10121-6

[9] Mallick RK, et al. *"A Critical Review of the Current Collector for PDMFCs"*. J Power Sources. 2015;285:2015 510–29.

[10] Yuan W, et al., 2013, *"Corrosion Behavior of Porous Metal Fiber-Sintered Felt in Both Simulated and Practical Environments of a DMFC"*, ISSN 0010-9312 (print), 1938-159X NACE International (2013)

[11] Hashemi R, Yousefi S, Faraji M. Experimental Studying of the Effect of Active Area on the Performance of PDMFC. Berlin Heidelberg: Springer-Verlag; 2015. 10.1007/s11581-015-1479-y. International Journal of Ionics.

[12] Yanga WM, Chou SK, Shu C. Effect of Current-collector Structure on Performance of Passive Micro Direct Methanol Fuel Cell. J Power Sources. 2007;164(2):10 February 2007 549–54. https://doi.org/10.1016/j.jpowsour.2006.11.014.

[13] Dohle H, Mergel J, Stolten D. Heat and Power Management of a Direct-methanol-fuel-cell (DMFC) System. J Power Sources. 2002;111(2):23 September 2002 268–82.

[14] Huang J, Baird DG, McGrath JE. Development of Fuel Cell Bipolar Plates from Graphite Filled Wet-lay Thermoplastic Composite Materials. J Power Sources. 2005;150:110–19. doi: 10.1016/j.jpowsour.2005.02.074.

[15] Abraham BG, Chetty R. Design and Fabrication of a Quick-fit Architecture Air Breathing Direct Methanol Fuel Cell. Int JHydrogen Energy. 2021;46(9):3 February 2021 6845–56.

[16] Song SQ, Liang ZX, Zhou WJ, Sun GQ, Xin Q, Stergiopoulos V, Tsiakaras P. Direct Methanol Fuel Cells: The Effect of Electrode Fabrication Procedure on MEAs Structural Properties and Cell Performance. J Power Sources. 2005;145(2):495–501. doi: 10.1016/j.jpowsour.2005.02.069.

[17] Cha H-C, Chen C-Y, $1-yuanshiu. Investigation on the Durability of Direct Methanol Fuel Cells. J Power Sources. 2009;192(2):451–56. doi: 10.1016/j.jpowsour.2009.03.028.

[18] Kuan Y-D, Lee S-M, Sung M-F. Development of a Direct Methanol Fuel Cell with Lightweight Disc Type Current Collectors. Energies. 2014;2014 7:3136–47. doi: 10.3390/en7053136.

[19] Raghavaiah NV, Naga Srinivasulu G. Experimental Investigations on Performance of PDMFC Using Nickel-201 and Brass Current Collector Materials. Chem Pap. 2022;76:4581–95.

[20] Raghavaiah NV, Srinivasulu G. "Analysis and Experimental Investigation on PDMFC Current Collectors with Taper Cylindrical Openings. Fuel Cells. 2022;22:169.

Nomenclature

PDMFC	Passive direct methanol fuel cell
CD	Current density
CC	Current collector
PD	Power density
Ni-201	Nickel with Grade 201composition
SS-316L	Stainless steel with Grade 316L composition
TCO	Taper cylindrical openings
UCO	Uniform cylindrical openings
MEA	Membrane Electrode Assembly in the Nomenclature

Ahsan Ul Haq and Suresh Kumar Reddy Narala*

In-Plane Dynamic Crushing Behavior of Hierarchical Honeycomb Sandwich Panel Under Blast Loading: A Finite Element Study

Abstract: Achieving superior shock mitigation and energy absorption under high impulsive blast loading, yet maintaining lightweight design feature, is still a challenge for defense scientific community. Hierarchical honeycombs, as lightweight and energy-absorbing materials, are being widely investigated owing to their improved structural stress and stiffness and therefore applying such lightweight structures for the purpose of blast mitigation could be a favorable approach. This study proposes the use of a hierarchical honeycomb core to enhance shock mitigation performance by increasing energy absorption and reducing back-sheet deflection. To assess the behavior of the hierarchical honeycomb sandwich panel (HHSP), air blast simulations are conducted using the ConWep code in Abaqus/Explicit. The in-plane crushing performance of the HHSP has been compared with a solid plate having the same areal density. The sandwich panel is subjected to various TNT loads at different stand-off distances. HHSP exhibits interesting in-plane crushing behavior as they effectively adapt to shock loading by gradually drawing material into the locally loaded region, thereby enhancing shock mitigation capacity. In contrast, traditional honeycomb panels undergo plastic deformation without localized stiffness enhancement. Against a 1 kg load, front-sheet bending followed by progressive core buckling is reported. While significant increase in cell wall buckling and core densification is reported against a 3 kg load. Moreover, the HHSP absorbs 27% more energy and experiences 19% less back-sheet deflection when compared with its monolithic counterpart. These findings provide a reliable basis for designing sandwich panels with hierarchical honeycomb cores to achieve enhanced blast resistance.

Keywords: Hierarchical core, sandwich panels, protective armor, air-blast loading, energy absorption

*Corresponding author: Suresh Kumar Reddy Narala, Department of Mechanical Engineering, BITS Pilani, Hyderabad Campus, Hyderabad 500078, Telangana, India, e-mail: nskreddy@hyderabad.bits-pilani.ac.in

Ahsan Ul Haq, Department of Mechanical Engineering, BITS Pilani, Hyderabad Campus, Hyderabad 500078, Telangana, India, e-mail: p20190449@hyderabad.bits-pilani.ac.in

https://doi.org/10.1515/9783111515717-008

1 Introduction

Traditional defense installations depend mostly on heavy metallic slabs to protect against the growing threat of explosions. Unfortunately, these measures significantly impede the functionality and mobility of equipments like aircraft and vehicles [1]. Protecting civilians from sudden explosions stands as a paramount objective for the scientific community engaged in defense industries [2]. Therefore, there is an immediate requirement for innovative and compact replacements that can offer superior blast mitigation. In recent times, honeycomb sandwich panels (HSPs) have emerged as a promising solution as they can effectively disperse shock impulses within the structure, consequently reducing pressure intensity on the other end of the panel. They are known for their lightweight nature, impressive stiffness, and exceptional energy absorption capabilities across various scenarios. Despite their lightweight nature, they exhibit remarkable resistance when subjected to high-impact dynamic loads [3].

It is widely recognized that achieving such remarkable mechanical properties is based on the structural topology. The core material is typically composed of aluminum, steel, or titanium, and it exhibits either a periodic or random cellular structure. Consequently, various honeycomb configurations, including hexagonal [1, 4], auxetic core [1, 3, 4], and Kagome [5], have been proposed and well investigated. Among these, the HSP is widely accepted due to its favorable mechanical characteristics and its prevalence in the natural world. Numerous studies have been conducted to explore both the out-of-plane and in-plane dynamic behaviors of honeycomb structures. Wang et al. [4] studied the out-of-plane crash-resistant behavior of six different cores, including hexagonal honeycomb. Hu et al. [6] examined the in-plane crashworthiness of auxetic honeycomb, focusing on energy absorption. Their findings revealed an increase in crushing strength with impact velocity. Liu et al. [7] investigated the in-plane crushing response of graded AHSP, utilizing ABAQUS to explore the impact of foil thickness and impact velocity.

Recently, a number of researchers have been directed toward assessing the blast performance of HSPs. Yan et al. [3] revealed that auxetic HSPs exhibit superior blast mitigation by reducing the early pressure pulse in contrast to other sandwich panels. Liu et al. [8] established that honeycomb panels filled with weaker tubes exhibit a significant enhancement in blast performance compared to foam panels. Dharmasena et al. [9] conducted experimental and numerical investigations, showcasing that HSPs substantially surpass monolithic solids of similar areal density. These HSPs exhibited smaller back-sheet deformation and the ability to absorb higher energy. Moreover, various factors influencing the blast resistance of HSPs, including core height [10], front sheet thickness, cell size [11], and foil thickness [12], have been thoroughly examined.

The advent of additive manufacturing (AM) has opened up new possibilities for creating intricate geometries within the crushable core, a task that would be challenging using traditional subtractive manufacturing methods. Recent research has illustrated the viability of AM parts in applications related to energy absorption, considering quasi-static as well as impact loading conditions [1, 13, 14]. Shape-memory honeycombs have been a recent innovation, produced through 4D printing technology, enabling the creation of reversible energy absorbers [15, 16]. Moreover, shape optimization [17] and topology optimization [18] techniques have been applied to design lightweight lattice structures featuring novel unit cell topologies. However, the design concepts mentioned above are constrained to single-order cellular honeycombs, which suffer from reduced in-plane stiffness, limiting their utility. To enhance the in-plane stiffness of low-density honeycombs, researchers have turned to nature for inspiration. One solution involves the concept of hierarchical configurations.

This study considers the advantages of hexagonal structures and takes a step further by introducing the concepts of hierarchical configuration, drawing inspiration from nature. Therefore, hexagonal honeycombs with hexagonal sub-structures are proposed, aimed at enhancing their shock-mitigating properties. The in-plane dynamic behavior of a high-ductility stainless steel sandwich panel featuring an AA 3003 hierarchical honeycomb is investigated. Initially, a finite element (FE) model of the panel is created in ABAQUS. Air blast simulations are conducted to analyze the in-plane crashworthiness of the honeycomb. It is compared to a monolithic solid plate under similar conditions. Furthermore, the influence of a TNT explosive charge on the crushing behavior of the honeycomb sandwich panel at various stand-off distances is examined and discussed in subsequent sections.

2 Computational Approach

In this study, three-dimensional FE analysis of the two-scale hierarchical honeycomb-cored sandwich panel under air blast is performed using ABAQUS/Explicit. This software is chosen for its high reliability, particularly in analyzing complex nonlinear dynamic problems [7].

2.1 Finite Element (FE) Model

A three-dimensional, two-scale hierarchical honeycomb sandwich panel measuring 300 mm × 300 mm × 50 mm was modeled. The hierarchical honeycomb core possesses a height of 40 mm and a wall thickness of 0.5 mm. The front and back sheets of the panel have a thickness of 5 mm. To optimize computational efficiency and reduce time, only one quarter of the sandwich panel is included in the model, considering the symmetry and loading conditions. Figure 1 illustrates the unit cell configuration, while Fig. 2 depicts one quarter of the numerically developed model.

2.2 Material Properties

For the sandwich panel, which comprises both the face sheets and the honeycomb core, high-ductility stainless steel has been selected for the front and back sheets due to its favorable properties, while AA 3003 is defined for the hierarchical honeycomb core. Many austenitic stainless steels exhibit a combination of these desirable characteristics [9]. Despite its relatively higher mass, high-ductility stainless steel demonstrates exceptional behavior and maintains its properties even when subjected to severe impacts [19, 20]. To accurately account for the strain rate effects crucial in the dynamic failure of energy-absorbing materials, the Johnson-Cook (J-C) plasticity model is employed. The mechanical properties of high-ductility stainless steel (SS) and AA 3003 [4, 9, 19] along with their corresponding J-C constants, are used in this work.

The static yield stress in the J-C model can be given as follows:

$$\sigma_y = \left(A + B\bar{\varepsilon}^{p^n}\right)\left(1 + C\ln\dot{\varepsilon}^*\right)\left(1 - T^{*m}\right) \tag{1}$$

where $\bar{\varepsilon}^{p^n}$ stands for the measure of non-elastic deformation and $\dot{\varepsilon}^*$ signifies the rate at which this deformation occurs. m serves as the thermal softening exponent. Material-specific constants A, B, C, and n, which are empirically determined, define how the material behaves under various conditions.

2.3 Interaction, Boundary Constraints, and Mesh details

The surface-to-surface contact algorithm within the interaction module was applied between the honeycomb and sheets, and a general contact condition with a dynamic explicit setting is established, featuring a coefficient of friction as 0.2. The blast scenario is defined using the ConWep technique in ABAQUS/Explicit [7]. Symmetrical boundary conditions in the X and Y axes are imposed on two edges, as depicted in Fig. 1b. The remaining two edges of the deformable sandwich panel are fully constrained ($U_x = U_y = U_z = U_{xy} = U_{yz} = U_{zx} = 0$). The front and back sheets are represented using eight-node linear brick (C3D8R) elements, while the core utilizes four-node shell (S4R) elements, as illustrated in Fig. 3. Each sheet comprises 34,162 nodes and 23,120 elements, while the core consists of 131,056 S4R elements. Following mesh convergence studies (as depicted in Fig. 4), an element size of 0.5 mm is selected.

3 Results and Discussions

A set of air blast simulations were carried out to explore the in-plane blast resistance of a two-scale hierarchical honeycomb-cored sandwich panel. To assess the structural

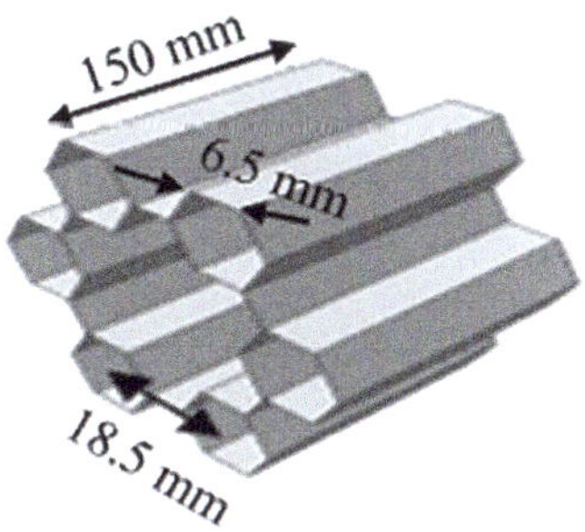

Fig. 1: Unit cell geometry of the hierarchical honeycomb.

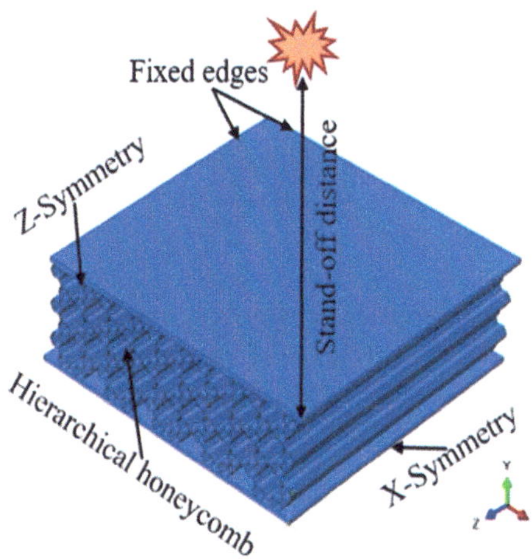

Fig. 2: Developed model of the sandwich panel.

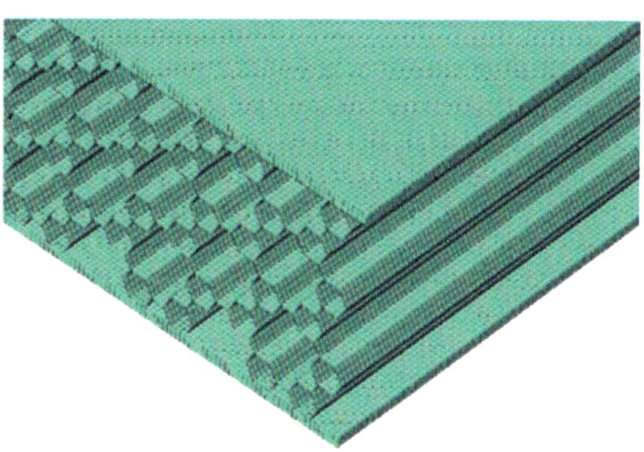

Fig. 3: Mesh details used for air blast studies.

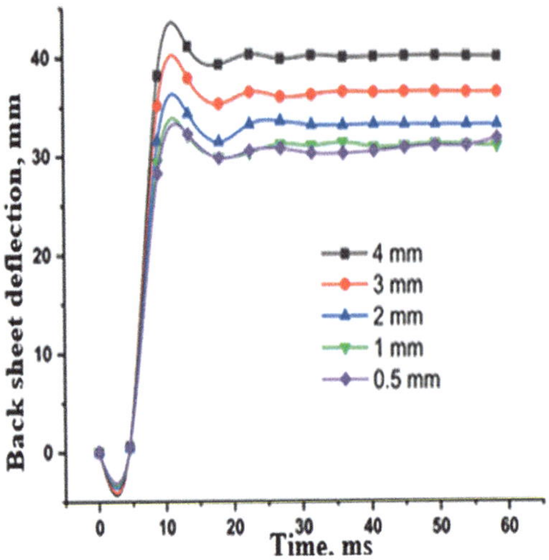

Fig. 4: Back-sheet deflections obtained for different sizes of mesh during mesh sensitivity studies.

response, a monolithic solid plate having the same areal density was employed for comparative analysis against the sandwich panel. The influence of varying TNT explosive quantities, such as 1, 2, and 3 kg, on the dynamic behavior of the honeycomb sandwich panel has been analyzed.

3.1 Pressure Distribution Using ConWep Technique

The ConWep blast simulation technique was employed to compute the spatial pressure distribution and impulse loading acting on the panel surface. This technique takes into account factors such as the stand-off distance, TNT charge, and the surface area of the panel. While it assumes a spherical charge, it provides a reasonably accurate approximation of the pressure profile [9]. Figure 5 illustrates the reflected pressure and the history of impulse loads experienced by the sandwich plate when a 1 kg TNT explosive is detonated 100 mm away from the target.

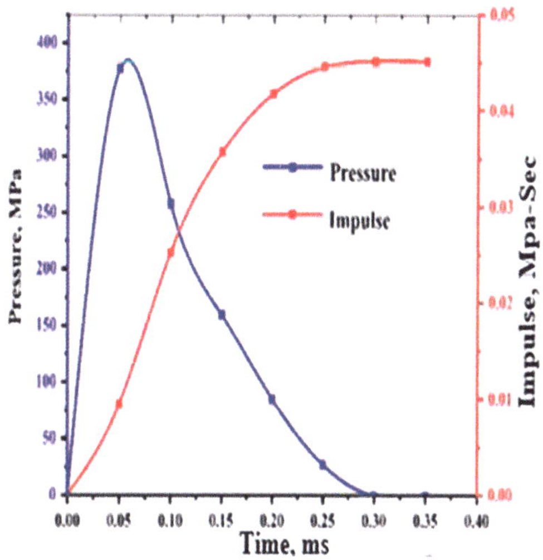

Fig. 5: Time histories of pressure and impulse against 1 kg TNT.

3.2 Comparison with Monolithic Solid Plate

The developed model is benchmarked against a monolithic solid plate with an equivalent areal density. In this comparison, a solid plate having 300 mm × 300 mm × 6 mm dimensions is chosen for blast loading, subjected to the same environment. To ensure a fair and balanced evaluation between the two, the mass of the solid plate is adjusted to approximately the same as that of the two-scale hierarchical honeycomb core [3]. In Fig. 6, the deformation contour of the equivalent solid plate can be observed, which exhibits a substantial deformation of 94.5 mm. To provide a comparative perspective, Fig. 6 displays the back sheet deflections in the hierarchical honeycomb sandwich panel, revealing a considerably reduced displacement of only 55.1 mm. This reduction emphasizes the superior performance of the HHSP in withstanding severe air blast conditions, especially when subjected to a 3 kg TNT explosive at a stand-off distance of 300 mm. Particularly, there is a substantial reduction in displacement, with the center of the front sheet experiencing a displacement of 76.4 mm and the back sheet showing a displacement of 55.1 mm, as depicted in Fig. 7.

In Fig. 8, the progressive deformation modes of a sandwich panel can be observed at different instances of time. Additionally, the fully deformed sandwich panel under 3 kg TNT and a 300 mm stand-off distance, along with an enlarged view of the in-plane buckling and crushing of the hierarchical honeycomb core, is shown in Fig. 9. The greater deflections observed in the front sheet can be attributed to the local stiffness at the center of the sheet [11]. In contrast, the smaller displacements in the back sheet are a result of the high flexural rigidity within the unit cells of the two-scale hierarchical honeycomb core.

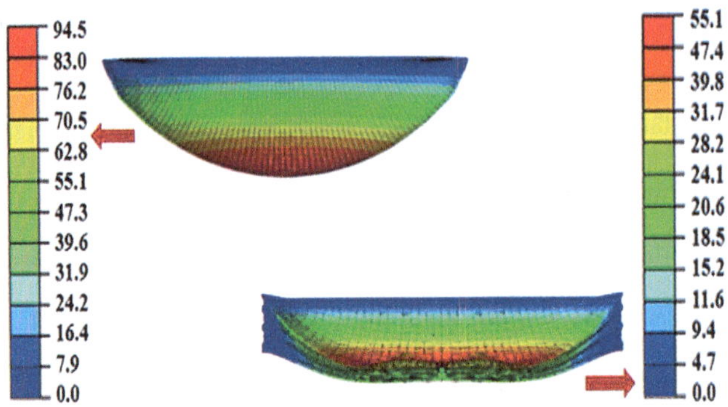

Fig. 6: Fully deformed contours of (a) solid plate and (b) sandwich panel under 3 kg TNT at 300 mm distance.

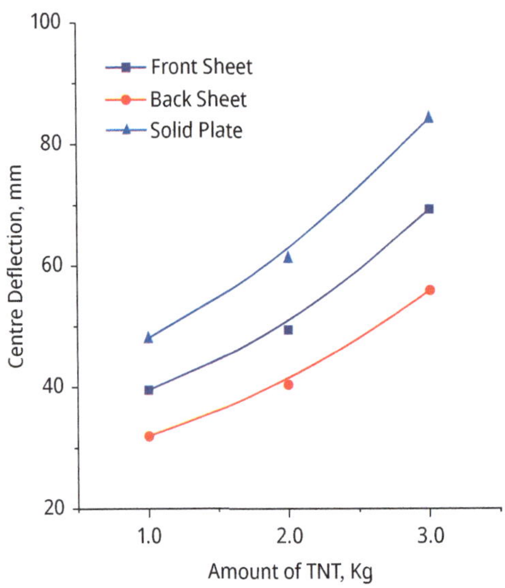

Fig. 7: Variation of center deflections for different parts with the explosive.

3.3 Energy Absorption Performance

The structural energy profile of the HHSP is illustrated in Fig. 10. Remarkably, it recorded values as high as 12.8 KJ, 11.95 KJ, and 4.1 KJ for total energy, internal energy, and kinetic energy, respectively. The results indicate that significant buckling and crushing occur within the hierarchical honeycomb core when exposed to an increasing quantity of explosives. For example, a maximum compressive strain of 0.74 was recorded for the case of 3 kg TNT. Notably, web crushing within the core happens rapidly, typically within a

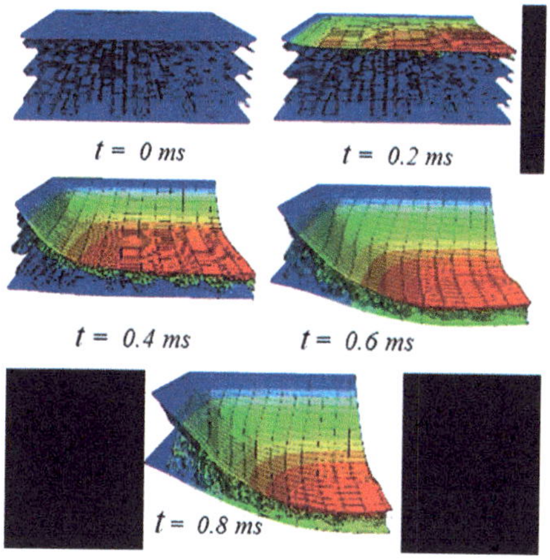

Fig. 8: Progressive deformation modes of a hierarchical honeycomb sandwich panel at different instances of time.

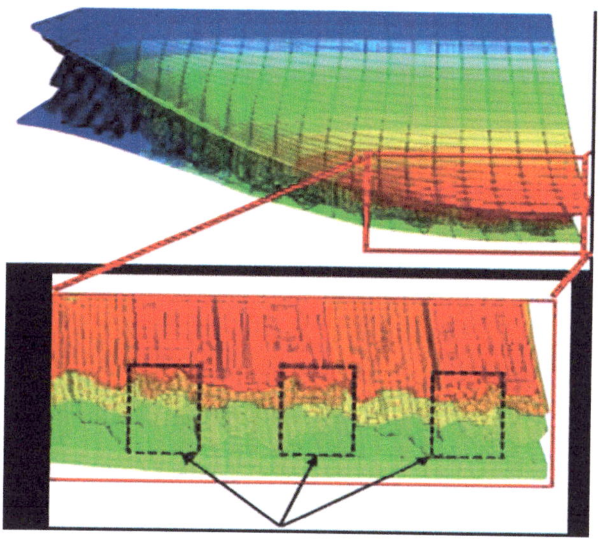

Fig. 9: Deformed sandwich panel with an enlarged view of in-plane buckling and crushing of the hierarchical honeycomb core.

time frame of less than 0.9 ms, followed by a spring-back effect [12]. The kinetic energy distribution between the pair of sheets and the honeycomb is depicted in Fig. 11. Front and back sheets exhibit nearly closer kinetic energy profiles over time. Interestingly, the core carries approximately twice the kinetic energy compared to each face sheet. Given that all three parts reach their maximum kinetic energy at almost the same time, suggesting a synchronized response of each component during the explosive event.

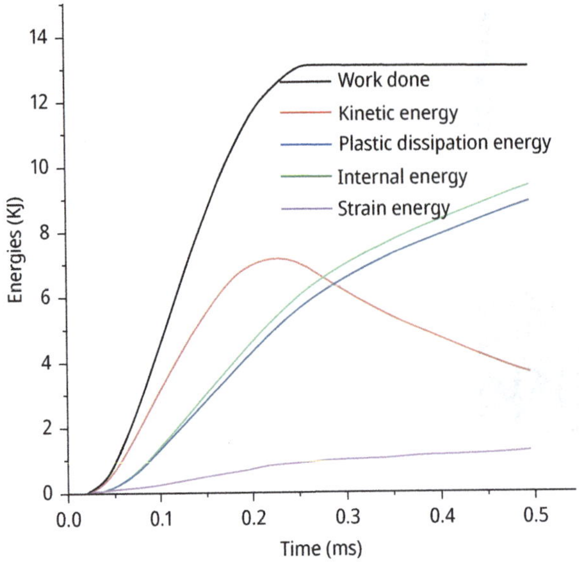

Fig. 10: Structural energy profiles of a sandwich panel as a function of time.

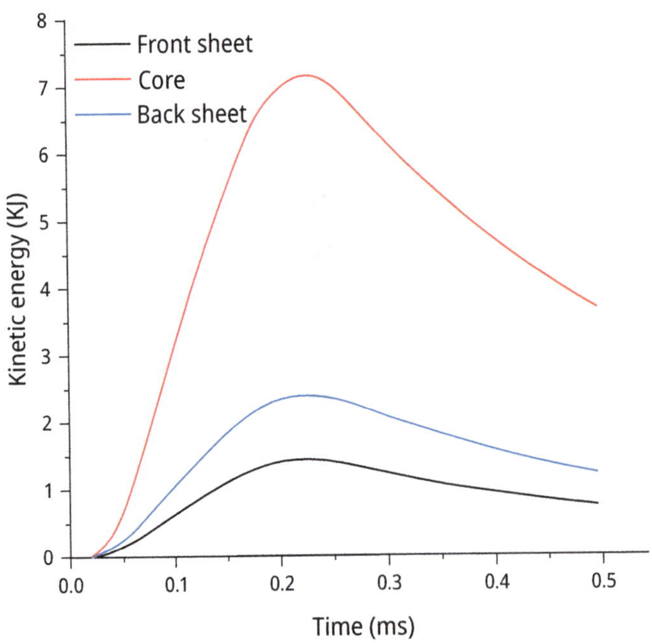

Fig. 11: Distribution of kinetic energy for different parts over time.

3.4 Influence of TNT and Stand-Off Distance

In Fig. 12, a graph illustrating plastic dissipation energy versus time is presented for different TNT charges. These results demonstrate that sandwich panels equipped with a two-scale hierarchical honeycomb core possess the capability to efficiently convert a substantial quantity of impact energy into plastic dissipation energy by way of face sheet bending, followed by buckling and crushing of the hierarchical core. The response of back sheet deflection under various TNT quantities and stand-off distances is presented in Fig. 13. Against a blast load of 3 kg TNT, the hierarchical honeycomb core experiences complete crushing, accompanied by significant bending of the back sheet. Furthermore, small debonding of the striking sheet is observed against a TNT load of 3 kg, particularly at a close distance of 100 mm. As the safe distance increases from 100 to 300 mm, a consistent drop in back sheet deformations is noted. However, it is noteworthy that sandwich panels equipped with the hierarchical honeycomb core exhibit approximately 18% lower deformations compared to those with the hexagonal honeycomb cores.

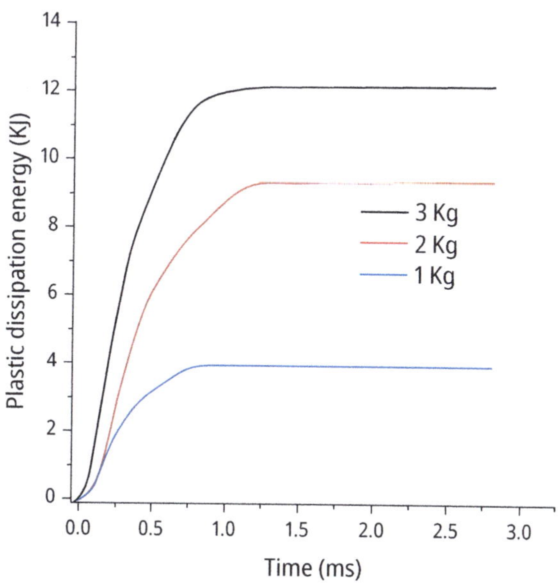

Fig. 12: Representation of plastic dissipation energy against different quantities of explosives.

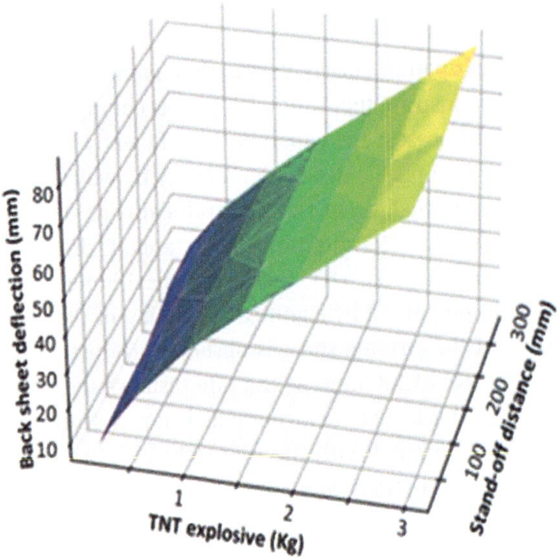

Fig. 13: Variation of back sheet displacements for different TNT and stand-off distances.

4 Conclusions

FE simulations were performed to investigate the in-plane crushing behavior of HHSPs. The influence of safe distance and explosive quantity on both maximum deflection and energy absorption capacity was investigated, resulting in the following conclusions:

- Under specified loading conditions, it is evident that the high-ductility stainless steel sandwich panel featuring a two-scale hierarchical honeycomb core outperforms both its monolithic equivalent and commercially available aluminum honeycomb sandwich panels in terms of blast resistance.
- The air blast loading simulations on the sandwich panels successfully captured various deformation mechanisms, including the stretching and bending of the face sheets, as well as the in-plane buckling and crushing of the hierarchical honeycomb core.
- The well-balanced two-scale hierarchical honeycomb sandwich panel configuration proved to be an innovative and optimal design, effectively dissipating the exploded impact energy while exerting lower forces at the other end.
- The study revealed that sandwich panel with hierarchical honeycomb cores demonstrates a 27% higher energy absorption capacity and an 18% reduction in deflection compared to those with hexagonal honeycomb cores. Furthermore, the deformation of the back sheet was 35% lower compared to the monolithic counterpart.

References

[1] Yan Z, Liu Y, Yan J, Wang B, Bai F, Shi Z, Huang F. Anti-blast Performance of 3D-printed Sandwich Panels with Auxetic Hexagonal and Regular Hexagonal Honeycomb Cores. Eng Struct. 2022 Dec 1;272:114996.

[2] Haq AU, Reddy NS. A Brief Review on Various High Energy Absorbing Materials. Mater Today Proc. 2021 Jan 1;38:3198–204.

[3] Ul Haq A, Gunashekar G, Narala SK. The Dynamic Response of AuxHex and Star-reentrant Honeycomb Cored Sandwich Panels Subject to Blast Loading. Arabian J Sci Eng. 2023 Sep;48 (9):11755–71.

[4] Li X, Kang R, Li C, Zhang Z, Zhao Z, Lu TJ. Dynamic Responses of Ultralight All-metallic Honeycomb Sandwich Panels under Fully Confined Blast Loading. Compos Struct. 2023 May 1;311:116791.

[5] Gu H, Pavier M, Shterenlikht A. Experimental Study of Modulus, Strength and Toughness of 2D Triangular Lattices. Int J Solids Struct. 2018 Nov 1;152:207–16.

[6] Tan HL, He ZC, Kx L, Li E, Cheng AG, Xu B. In-plane Crashworthiness of Re-entrant Hierarchical Honeycombs with Negative Poisson's Ratio. Compos Struct. 2019 Dec 1;229:111415.

[7] Liu H, Zhang ET, Wang G, Ng BF. In-plane Crushing Behavior and Energy Absorption of a Novel Graded Honeycomb from Hierarchical Architecture. Int J Mech Sci. 2022 May 1;221:107202.

[8] Liu J, Wang Z, Hui D. Blast Resistance and Parametric Study of Sandwich Structure Consisting of Honeycomb Core Filled with Circular Metallic Tubes. Compos Part B Eng. 2018 Jul 15;145:261–69.

[9] Dharmasena KP, Wadley HN, Xue Z, Hutchinson JW. Mechanical Response of Metallic Honeycomb Sandwich Panel Structures to High-intensity Dynamic Loading. Int J Impact Eng. 2008 Sep 1;35 (9):1063–74.

[10] Chi Y, Langdon GS, Nurick GN. The Influence of Core Height and Face Plate Thickness on the Response of Honeycomb Sandwich Panels Subjected to Blast Loading. Mater Des. 2010 Apr 1;31 (4):1887–99.

[11] Li X, Zhang P, Wang Z, Wu G, Zhao L. Dynamic Behavior of Aluminum Honeycomb Sandwich Panels under Air Blast: Experiment and Numerical Analysis. Compos Struct. 2014 Feb 1;108:1001–08.

[12] Meng Y, Lin Y, Zhang Y, Li X. Study on the Dynamic Response of Combined Honeycomb Structure under Blast Loading. Thin-Walled Struct. 2020 Dec 1;157:107082.

[13] Wei Y, Zhang C, Yuan Y, Chen P, Huang C, Li J, Yuan M. Blast Response of Additive Manufactured Ti–6Al–4V Sandwich Panels. Int J Impact Eng. 2023 Apr 1;176:104553.

[14] Xiao L, Song W. Additively-manufactured Functionally Graded Ti-6Al-4V Lattice Structures with High Strength under Static and Dynamic Loading: Experiments. Int J Impact Eng. 2018 Jan 1;111:255–72.

[15] Bodaghi M, Liao WH. 4D Printed Tunable Mechanical Metamaterials with Shape Memory Operations. Smart Mater Struct. 2019 Mar 20;28(4):045019.

[16] Bodaghi M, Serjouei A, Zolfagharian A, Fotouhi M, Rahman H, Durand D. Reversible Energy Absorbing Meta-sandwiches by FDM 4D Printing. Int J Mech Sci. 2020 May 1;173:105451.

[17] Bertolino G, Montemurro M, De Pasquale G. Multi-scale Shape Optimisation of Lattice Structures: An Evolutionary-based Approach. Int J Interact Des Manuf (Ijidem). 2019 Dec;13(4):1565–78.

[18] Montemurro M, Bertolino G, Roiné T. A General Multi-scale Topology Optimisation Method for Lightweight Lattice Structures Obtained through Additive Manufacturing Technology. Compos Struct. 2021 Feb 15;258:113360.

[19] Haq AU, Kavit AK, Rao T, Buddi T, Baloji D, Satyanarayana K, Singh SK. Evaluation and Optimization of Material Properties of ASS 316L at Elevated Temperatures Using Response Surface Methodology. Mater Today Proc. 2019 Jan 1;18:4589–97.

[20] Haq AU, Narala SK. Dynamic Response of Honeycomb Cored Sandwich Panels under High Velocity Impact: A Numerical Study. Mater Today Proc. 2023 Mar 22.

Katika Harikrishna*, Maharajan Joseph Davidson, Gottim Dhanush Reddy, Poorna Jagadeesh, Vasista Kadali, and Shaik Naveed Ahammed

Utilizing the U-Net Image Machine Learning Segmentation Model for Automated Analysis of Grain Sizes in Metallographic Microstructures

Abstract: Grain size analysis is a critical aspect of the study of metallographic microstructures, directly influencing material properties. Accurate determination of grain size necessitates precise image segmentation techniques, encompassing both image processing methods and emerging machine learning approaches. In this study, image processing techniques were employed to establish boundaries and label micrographs obtained after an elevated-temperature plastic deformation process. The image simplification process involves converting RGB microstructure images to gray scale, followed by the application of median blur to enhance image smoothness and segmentation accuracy. The Otsu method, a thresholding technique, is employed to differentiate grains from boundaries, and erosion and dilation filters are utilized for grain separation. Additionally, the study introduces the use of the U-Net model to automate the image segmentation process for complex optical images, even with limited training data. Experimental results showcase the notable advantages of this approach, particularly when dealing with optical images that exhibit low contrast, indistinct boundaries, and complex structures. By comparing with manual methods, the U-Net model achieves a significantly reduced average relative error of approximately 0.583%. Furthermore, grain size calculations based on the segmentation, adhering to the standards set by ASTM, yield satisfactory results.

Keywords: U-Net model, Image segmentation, Image processing methods, machine learning

*Corresponding author: **Dr. Katika Harikrishna**, Department of Mechanical Engineering, National Institute of Technology, Warangal 506004, Telangana, India; Department of Mechanical Engineering, Shri Vishnu Engineering College for Women, Bhimavaram 534202, Andhra Pradesh, India, e-mail: kharikrishna@student.nitw.ac.in

Maharajan Joseph Davidson, Gottim Dhanush Reddy, Poorna Jagadeesh, Vasista Kadali, and Shaik Naveed Ahammed, Department of Mechanical Engineering, National Institute of Technology, Warangal 506004, Telangana, India

https://doi.org/10.1515/9783111515717-009

1 Introduction

The microstructure of materials provides essential insights into important characteristics such as mechanical strength, abrasion resistance, surface texture, resistance to corrosion, coating quality, and ease of machining [1]. Factors such as the introduction of alloying elements, manufacturing methods, and post-processing methods play a pivotal role in bringing about changes in microstructure [2]. Various techniques have been used in the past to approach segmenting an image, the method of splitting an image into sections having comparable characteristics. These segmentation methods include region [3], threshold [4], graph theory [5], edge detection [6], and energy functions [7]. Threshold-based segmentation involves categorizing image pixels into different groups based on feature thresholds. It encompasses techniques such as histogram threshold segmentation [8], iterative determination [9], inter-class variance threshold [10], and dynamic threshold [11]. Region-based segmentation, on the other hand, divides the image into distinct regions based on a similarity criterion. Edge-based segmentation methods like the Kirsch [12], Sobel [13], Canny [14], and Log [15] are effective for edge detection but less suitable for complex data or micrographs with significant noise. Segmentation methods that employ upsampling or deconvolution techniques include SegNet [16], FCN (fully convolutional network) [17], and U-Net [18]. The FCN, proposed by Jonathan Long, is specifically designed for semantic segmentation at the pixel level. U-Net, which builds upon FCN, performs semantic segmentation and has shown superior performance in the ISBI challenge with minimal training data. SegNet, another pixel-level classification network, features an encoder-decoder architecture that uses pooled indices for information transfer. For the extraction of road areas, ResU-Net [19], a segmentation neural network, blends residual learning with U-Net architecture. It leverages residual units and exhibits similarities to U-Net in its design. While these segmentation methods have been widely used for metallographic images, their effectiveness is limited when dealing with complex images such as optical images. One reason for this limitation is the requirement for a huge number of labeled training images in machine learning-based methods. Additionally, complex images like optical images often have a significant background portion, making the task of segmenting the grain boundaries, which occupy a small pixel proportion, challenging. As a result, existing loss functions struggle to optimize the network effectively, leading to significant deviations in the measured grain size. Since metal performance is closely tied to the average grain size, such deviations can severely affect the accuracy of predicted metal properties.

To deal with these problems, this chapter proposes a U-Net model for grain size measurement, specifically suited for complex images such as optical images. The U-Net model presented in this work demonstrates superior performance compared to other models, even with a small training dataset. The following is a summary of this chapter's contributions: firstly, the introduction of a U-Net model trained on a small dataset, which effectively segments complex metallographic images. Furthermore, the image segmentation results obtained are used to estimate the grain size using the planimetric method based on the ASTM standard (E112-12), yielding satisfactory results.

2 Materials and Methods

The researchers conducted a previous study to investigate the impact of magnesium (Mg) in the Al-5.6Zn system. The results demonstrated that the inclusion of 2% Mg in the alloy led to improved properties. The earlier study [20, 21] also provided a detailed description of the fabrication process by means of powder metallurgy (PM). The preforms, after sintering, underwent isothermal deformation with a strain rate of 0.005 s^{-1} applied at temperatures of 300 °C, 400 °C, and 500 °C until fracture occurred. As stated by Seetharam et al. [22], sintered preform experiences three distinct stages of compression. After undergoing compression, three specimens were chosen.

3 Results and Discussion

3.1 Data Collection

In this chapter, hot-deformed powder metallurgy optical micrographs were selected as dataset from the previous studies of authors [23]. To enhance the training dataset and increase its diversity, data augmentation techniques were applied. Specifically, operations such as rotation and mirroring were performed on the original and labeled images. Each original image resulted in three additional augmented images, effectively expanding the training dataset and providing the model with more variations of the microstructure images. The dataset was parted into the training set and the test set. The training set contained 36 optical images, and the test set consisted of 3 optical images.

3.2 Image Labeling

By using image processing, segmentation has been done to determine their boundaries and labeled our optical images. The process of image simplification includes altering images from RGB to gray scale. Median blur is used to smooth the micrograph and improve segmentation accuracy. The thresholding technique, specifically the Otsu method, is used to separate the foreground (boundaries) and background (grains), and erosion and dilation filters are applied for grain division. With enough data in this way, the network model's training effect could be enhanced, and it had strong resilience and invariance. Figure 1 shows the output images for training. The input images are available in the prior research conducted by Harikrishna et al. [2, 23, 24].

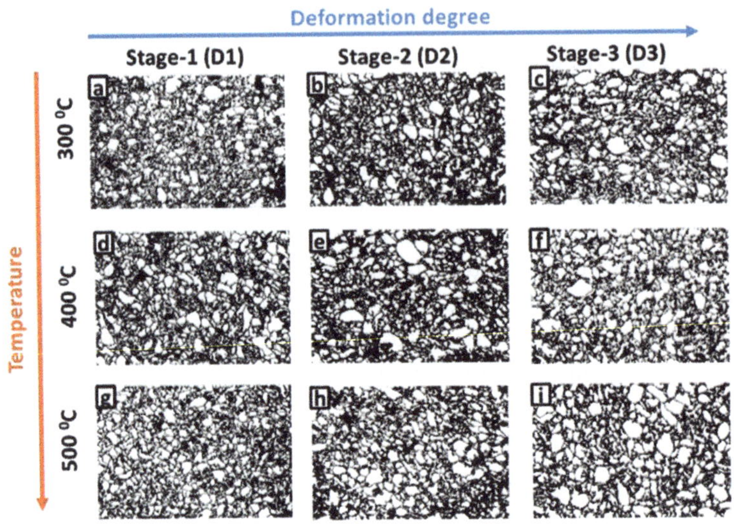

Fig. 1: Output optical images for the ML model.

3.3 The U-Net Model

The U-Net model is a powerful model extensively used for microstructure segmentation in optical images. It effectively identifies and classifies various regions within the image by leveraging its unique design. This architecture employs an encoder–decoder network to capture both global and local features of the microstructure. The encoder learns hierarchical representations through convolutional layers, while the decoder combines this information to generate a high-resolution segmentation map. Spatial details are preserved through skip connections between the corresponding encoder and decoder layers. Figure 2a illustrates the structure of the U-Net segmentation model. A U-Net model was proposed in this study to segment images, and satisfactory segmentation results were obtained using only thirty-six training images. The provided algorithm performs the segmentation of grains and boundaries in optical images using the fastai library. The flowchart of the employed U-Net model is presented in Fig. 2b. A brief description of the model is provided below.

Import necessary modules: The algorithm starts by importing the required modules, including the fastai.vision module for computer vision tasks and the matplotlib module for image visualization. Retrieve original image file paths: The code retrieves the file paths of the original images using the get_image_files function. Define the label function: The label_func function is defined to determine the corresponding labeled image path for a given original image path. Load codes for grain and background: The code loads the codes for grain and background from the "labels.txt" file using the np.loadtxt function. These codes will be used to map the labels during the segmentation process. Create SegmentationDataLoaders: The SegmentationDataLoaders class is used to create data loaders for segmentation. It takes the labeled and original image paths, along with other parameters such as batch size (bs), label function (label_func), and image resizing (item_tfms). Create a UNET learner: The unet_learner function is used to create a UNET model using the ResNet34 architecture. It takes the SegmentationDataLoaders (dls) and the desired backbone architecture (resnet34 in this case) as parameters. Fine-tune the UNET model: The fine_tune method is called on the UNET learner (learn) to fine-tune the model for 20 epochs. This step trains the model using the provided data loaders and updates the weights to improve segmentation performance. Overall, this code segment sets up the data loaders, creates a UNET model, and fine-tunes it for segmentation of grain and boundaries in optical images. After implementing the algorithm described earlier, the U-net model successfully performed image segmentation on the new images. The resulting image is presented in Fig. 3, while Fig. 4 showcases the segmented images generated by the U-net model. Presented model made as an app and can be available at https://huggingface.co/spaces/dhanushreddy29/microstructure-project, accessed on July 5, 2023. The ASTM grain size number was calculated by following the procedure given by Fatih Akkoyun and Ali Ercetin [25].

Tab. 1: Comparison of ASTM grain size measurement between manual method and U-Net model.

S. no.	ASTM number (manual)	ASTM number (U-Net)	Relative error (%)
1	24.33	24.30	0.123
2	23.87	23.93	0.25
3	23.64	23.69	0.21

When analyzing the findings, it is essential to recognize that manual methods are not infallible, as they can be influenced by subjectivity and bias. Thus, accurately enumerating these errors remains challenging. This study, therefore, adopts relative error as an indicator. The experimental investigations revealed that both automated measurement techniques and theoretical estimations returned comparable outcomes, as illustrated in Table 1. The analysis determined that the average relative error in the ASTM number was 0.194%. These results highlight the strong agreement between the experimental data and theoretical predictions. A separate study [26] compared to manual measurements and MiPar analysis; the Watershed Algorithm exhibited a relative error of approximately 6.66%, with a deviation of 0.3 μm relative to the average grain size of 4.5 μm.

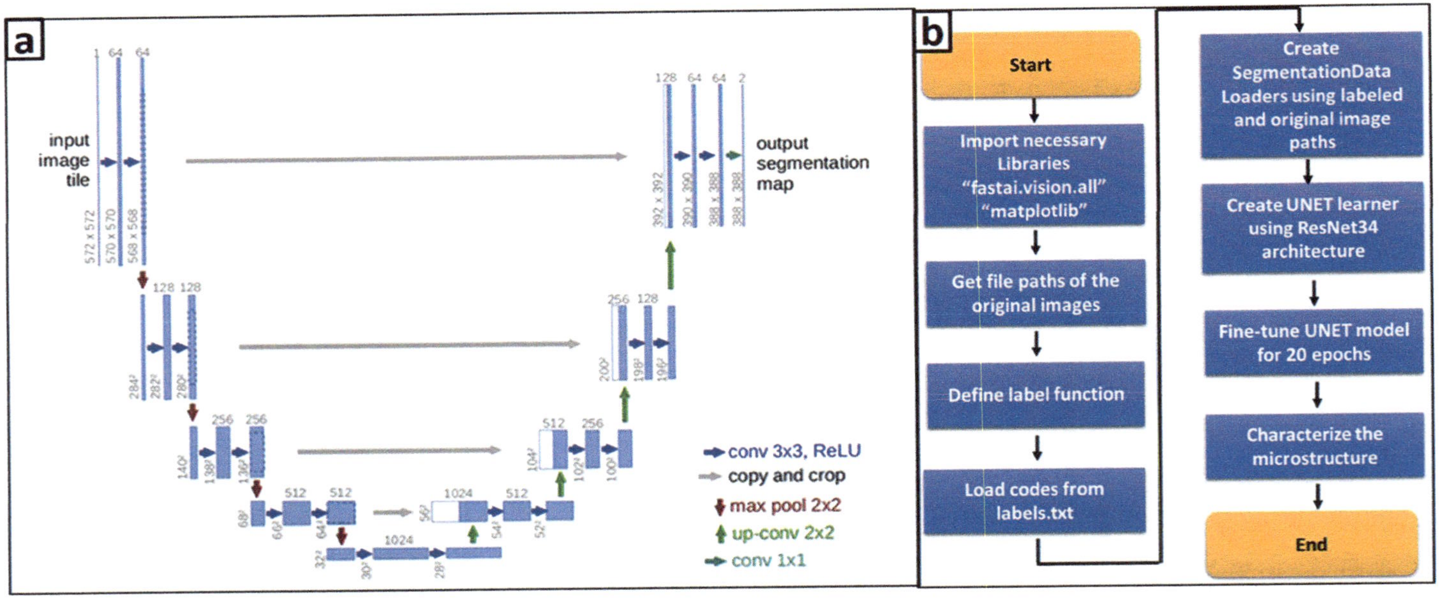

Fig. 2: (a) The U-Net model and (b) the structure of the U-Net segmentation model.

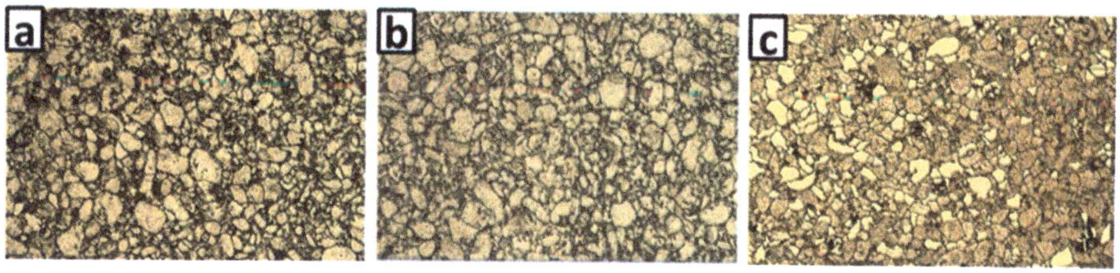

Fig. 3: Images used for testing the U-Net model.

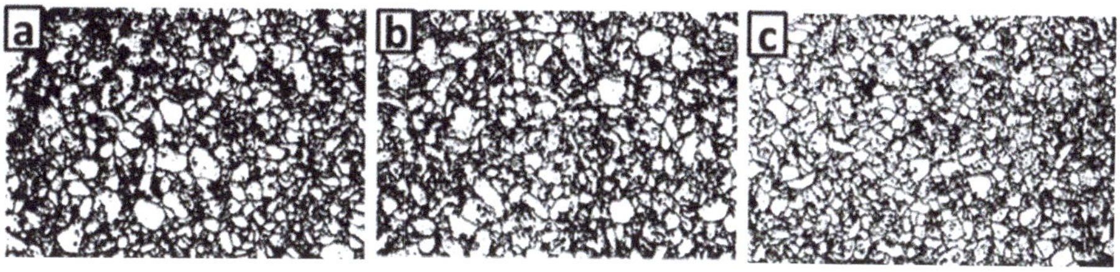

Fig. 4: Segmented images generated by the U-Net model.

4 Results and Discussion

The research effectively showcased the U-Net model's efficacy in segmenting microstructures in optical images and its ability to generate precise outcomes that are comparable to manual measurements. Notable findings from this study encompass the subsequent observations:

- Image processing techniques were applied to annotate and obtain boundaries of the metallographic images.
- The U-Net model achieved satisfactory segmentation results with just 36 training images.
- The U-Net model was compared with manual measurements of the ASTM grain size number. The relative error between the manual measurements and U-Net predictions was found to be very low, ranging from 0.123% to 0.25%.
  The calculation results showed consistent outcomes, reinforcing the reliability of the automated measurements.

References

[1] Nithin AM, Davidson MJ, Rao CSP. Mechanical and Metallurgical Studies on Thixoextruded Al-Si Alloys. J Mater Eng Perform. 2020;29(10):6378–89. doi: 10.1007/s11665-020-05143-8.

[2] Harikrishna K, Davidson MJ, Reddy GD. New Method for Microstructure Segmentation and Automatic Grain Size Determination Using Computer Vision Technology during the Hot Deformation of an Al-Zn-Mg Powder Metallurgy Alloy. J Mater Eng Perform. 2023. doi: 10.1007/s11665-023-09025-7.

[3] Kaganami HG, Beiji Z. Region-based Segmentation versus Edge Detection. IIH-MSP 2009 – 2009 5th Int Conf Intell Inf Hiding Multimed Signal Process. 2009;1217–21. doi: 10.1109/IIH-MSP.2009.13.

[4] Bhargavi K, Jyothi S. A Survey on Threshold Based Segmentation Technique in Image Processing. Int J Innov Res Dev. 2014;3(12):234–39. [Online]. Available https://www.researchgate.net/publication/309209325.

[5] Parihar VR, Thakur NV. Graph Theory Based Approach For Image Segmentation Using Wavelet Transform. Int J Image Process. 2014;8(5):255–77. [Online]. Available http://www.cscjournals.org/manuscript/Journals/IJIP/Volume8/Issue5/IJIP-904.pdf.

[6] Huang YC, Tung YS, Chen JC, Wang SW, Wu JL. An Adaptive Edge Detection Based Colorization Algorithm and Its Applications. Proc 13th ACM Int Conf Multimedia, MM 2005. 2005;351–54. doi: 10.1145/1101149.1101223.

[7] Bresson X, Vandergheynst P, Thiran JP. A Priori Information in Image Segmentation: Energy Functional Based on Shape Statistical Model and Image Information. IEEE Int Conf Image Process. 2003;3:425–28. doi: 10.1109/icip.2003.1247272.

[8] Tobias OJ, Seara R. Image Segmentation by Histogram Thresholding. IEEE Trans Image Process. 2002;11(12):1457–65. doi: 10.1109/TIP.2002.806231.

[9] Rezaiee-Pajand M, Entezami A, Shariatmadar H. An Iterative Order Determination Method for Time-series Modeling in Structural Health Monitoring. Adv Struct Eng. 2018;21(2):300–14. doi: 10.1177/1369433217717118.

[10] Arifin AZ, Asano A. Image Segmentation by Histogram Thresholding Using Hierarchical Cluster Analysis. Pattern Recognit Lett. 2006;27(13):1515–21. doi: 10.1016/j.patrec.2006.02.022.

[11] Yogarajah P, Condell J, Curran K, Cheddad A, McKevitt P. A Dynamic Threshold Approach for Skin Segmentation in Color Images. Proc – Int Conf Image Process ICIP. 2010;2225–28. doi: 10.1109/ICIP.2010.5652798.

[12] Decordier I, Papine A, Vande Loock K, Plas G, Soussaline F, Kirsch-Volders M. Automated Image Analysis of Micronuclei by IMSTAR for Biomonitoring. Mutagenesis. 2011;26(1):163–68. doi: 10.1093/mutage/geq063.

[13] Gao W, Yang L, Zhang X, Liu H. An Improved Sobel Edge Detection. Proc – 2010 3rd IEEE Int Conf Comput Sci Inf Technol ICCSIT 2010. 2010;5:67–71. doi: 10.1109/ICCSIT.2010.5563693.

[14] Eisenbach M, et al. How to Get Pavement Distress Detection Ready for Deep Learning? A Systematic Approach. Proc Int Jt Conf Neural Networks. 2017;2017-May(June 2021):2039–47. doi: 10.1109/IJCNN.2017.7966101.

[15] Marr D, Hildreth E. Theory of Edge Detection. Proc R Soc London – Biol Sci. 1980;207(1167):187–217. doi: 10.1098/rspb.1980.0020.

[16] Badrinarayanan V, Kendall A, Cipolla R. SegNet: A Deep Convolutional Encoder-Decoder Architecture for Image Segmentation. IEEE Trans Pattern Anal Mach Intell. 2017;39(12):2481–95. doi: 10.1109/TPAMI.2016.2644615.

[17] Long J, Shelhamer E, Darrell T. Fully Convolutional Networks for Semantic Segmentation. Intas Polivet. 2009;10(2):227–28.

[18] Brox T, Ronneberger O, Fischer P. U-Net: Convolutional Networks for Biomedical Image Segmentation Olaf. Lect Notes Comput Sci (Including Subser Lect Notes Artif Intell Lect Notes Bioinformatics). 2015;9351(Cvd).12–20. doi: 10.1007/978-3-319 24574 4.
[19] Zhang Z, Liu Q, Wang Y. Road Extraction by Deep Residual U-Net. IEEE Geosci Remote Sens Lett. 2018;15(5):749–53. doi: 10.1109/LGRS.2018.2802944.
[20] Krishna KH, Davidson MJ, Rajmohan KS, Bharat N. Influence of Incremental Pre-strain on Corrosion Behavior of Al-Zn-Mg Powder Metallurgy Alloy during Hot Forging. Proc Inst Mech Eng Part C J Mech Eng Sci. 2022;237(13):2990–3008. doi: 10.1177/09544062221143001.
[21] Hari Krishna K, Mj D, KS R, Nithin AM. Role of Pre-strain on the Corrosion Behaviour of Al-Zn-Mg P/M Alloy. Proc Inst Mech Eng Part L J Mater Des Appl. 2023;237(1):218–33. doi: 10.1177/14644207221109676.
[22] Seetharam R, Subbu SK, Davidson MJ. Hot Workability and Densification Behavior of Sintered Powder Metallurgy Al-B4C Preforms during Upsetting. J Manuf Process. 2017;28:309–18. doi: 10.1016/j.jmapro.2017.06.012.
[23] Harikrishna K, Davidson MJ, Seetharam R, Sundar S, Rajmohan KS. Investigating the Effects of Deformation-induced Densification on the Constitutive Behavior and Corrosion Resistance of Al–Zn–Mg Powder Metallurgy Alloy during Hot Deformation. Proc Inst Mech Eng Part L J Mater Des Appl. 2023. doi: 10.1177/14644207231207820.
[24] Harikrishna K, Davidson MJ, Reddy GD, Veera Venkata Nagaraju K. Overcoming Optical Image Challenges in Automatic Grain Size Measurement Using a Novel Computer Vision Algorithm Applied to Hot Deformation of Al-Zn-Mg Powder Metallurgy Alloy. Mater Lett. 2024;357 (November 2023):135743. doi: 10.1016/j.matlet.2023.135743.
[25] Ercetin A, et al. Image Processing of Mg-al-sn Alloy Microstructures for Determining Phase Ratios and Grain Size and Correction with Manual Measurement. Mater (Basel). 2021; 14(17). doi: 10.3390/ma14175095
[26] Campbell A, Murray P, Yakushina E, Marshall S, Ion W. New Methods for Automatic Quantification of Microstructural Features Using Digital Image Processing. Mater Des. 2018;141:395–406. doi: 10.1016/j.matdes.2017.12.049.

Gaddam Ashok, Pankaj Kumar*, and Ajay Kumar

Advancements in Banana Fiber-Reinforced Composites: Mechanical Properties, Sustainability, and Scientometric Analysis

Abstract: Banana fiber-reinforced composites (BFRCs) have gained attention for their eco-friendly and sustainable characteristics, with this study focusing on their mechanical and thermal properties to assess their potential for various applications. Alkali treatment of banana fibers significantly enhances their compatibility with polymer matrices, resulting in increased tensile strength, Young's modulus, and thermal stability. The composites, particularly with a 20% fiber reinforcement level, exhibited notable improvements in tensile, compressive, and impact strengths, making them a promising alternative to conventional materials for use in automotive, construction, and packaging industries. A scientometric analysis of digital fabrication technology from 2000 to 2023 revealed a substantial annual growth rate of 14.81%, with 232 publications and 483 research papers. The field has a robust research community of 2,342 authors, with 17.63% of research involving international collaboration and an average citation rate of 26.62 per paper, demonstrating significant impact. India has emerged as a leader in this domain with the highest number of publications and authors. This study provides valuable insights into BFRCs and the evolving landscape of digital fabrication technology, serving as a resource for future research.

Keywords: Banana fiber, natural fiber, composite material, scientometrics, mechanical properties, thermal properties, biocomposites, hybrid composites, reinforcement, sustainability, alkali treatment, fiber-matrix bonding, eco-friendly materials, automotive composites, biodegradability

***Corresponding author: Pankaj Kumar**, Department of Mechanical Engineering, SR University, Warangal 506371, Telangana, India, e-mail: pikupankaj82@gmail.com
Gaddam Ashok, Department of Mechanical Engineering, SR University, Warangal 506371, Telangana, India
Ajay Kumar, Department of Mechanical Engineering, School of Core Engineering, Faculty of Science Technology and Architecture, Manipal University Jaipur, Rajasthan, India

https://doi.org/10.1515/9783111515717-010

1 Introduction

1.1 Background

Over the past few decades, much attention has been paid to the utilization of natural fibers in the reinforcement of composite materials. This trend is attributed to the growing importance of using sustainable and environmentally friendly materials. Natural fiber composites can be considered as new generation substitutes for synthetic fiber composites owing to their renewability, biodegradability, low cost, and reasonably good mechanical properties. Among the different types of natural fibers, banana fibers are of increasing interest because of their abundance, acceptable mechanical properties, and utilization of agricultural residues. Banana fibers are obtained from the pseudo-stem of the banana plant (*Musa* species), which is widely grown in tropical and sub-tropical climates. It entails the removal of fibrous components from non-fibrous tissues using physical, chemical, or biological methods. Mechanical extraction is the most frequently used method owing to its ease of use and low cost. The final fibers have a high cellulose content that provides mechanical strength and a comparatively low density, which can be incorporated in lightweight composites.

1.2 Objectives and Significance of the Study

This study aims to comprehensively analyze key research trends and themes to understand the main areas of interest and innovation within the field. It quantifies research output and impact by examining publication metrics, providing insights into the productivity and influence of the research community. Additionally, this study will evaluate collaboration and authorship patterns to identify key contributors, networks, and potential areas for increased cooperation. By assessing journal influence and publication sources, it seeks to recognize primary venues for disseminating research and informing future publication strategies. Finally, it identifies research gaps and suggests future directions to guide new studies and address unresolved issues in the field.

The study of banana fiber-reinforced composites (BFRCs) through a scientometric review is significant, both academically and practically. By systematically analyzing the existing literature, this review advances knowledge by highlighting significant findings, methodologies, and technological advancements in BFRC research. It underscores the sustainability and environmental benefits of using natural, renewable resources such as banana fibers, which offer biodegradability and a reduced carbon footprint, aligning with global sustainability efforts. Additionally, this study guides future research by identifying gaps and suggesting directions, serving as a valuable resource for researchers, funding agencies, and policymakers. Furthermore, by understanding the properties and potential applications of BFRCs, this review contributes to the development of enhanced composite materials with improved mechanical properties, durability, and performance that are applicable in industries such as automotive, construction, and consumer goods.

2 Literature Review

Banana fiber reinforcement composites have been the subject of much research attention, mainly because of their low cost, renewability, and high specific strength and modulus. To turn banana fibers into clothes, banana plant stalks would otherwise be considered as waste products once bananas have been harvested [1]. These natural fibers are biodegradable and have a low density that does not exert much pressure on processing tools; they are suitable for several industries [2, 36–39]. Adding banana fibers has been reported to enhance the thermal stability, tensile strength, and impact strength of polymer matrices including polypropylene (PP), polylactic acid, and polyester [3]. For instance, PP banana fiber composites (PPBC) exhibit better thermo-oxidative stability and mechanical properties for construction and engineering applications [4]. Similarly, the use of banana fibers in concrete has been found to improve the compressive and tensile strengths, crack control, and spall resistance of ordinary cement concrete, where banana fibers can be termed as natural concrete addition [4]. Banana fiber composites can be hybrid composites in which the banana fiber is reinforced with other materials such as multi-walled carbon nanotubes and rice husk fillers, which improves the mechanical properties of the material such as hardness, tensile strength, and flexural strength. Moreover, it has been reported that the use of banana fibers in multi-layer systems with virgin PE increases the Young's modulus, but the impact strength remains an issue [5, 6]. It has been reported that the chemical treatment of banana fibers, especially with NaOH, enhances the fiber-matrix bond as well as the compatibility between the fiber and the matrix, which enhances the mechanical properties of the composite [7, 8]. Banana fiber composites are innumerous and can be used domestically in aircraft and automotive parts owing to their environmental friendliness and light weight [9]. Researchers have also studied the application of false banana fibers as reinforcement materials in polymer matrices owing to their attractive physical, mechanical, and thermal properties [9]. Collectively, the expanding body of research emphasizes the potential of BFRCs as a sustainable substitute for synthetic fiber composites, offering a combination of environmental advantages and improved material performance [10, 40–42].

The process resulted in a 30% reduction in hemicellulose, 25% reduction in lignin, and a 20% increase in cellulose content, thereby enhancing cellulose crystallinity [11, 12]. Banana fiber is considered a sustainable bioresource for eco-friendly applications owing to its mechanical and fracture behavior, energy-efficient extraction drives, and fabrication and characterization of bio-nanocomposites based on banana fiber [13–16].

Incorporating untreated and alkali-treated banana fibers into SEBS composites significantly influences the material properties. Untreated fibers generally exhibited higher tensile strength, except at 5% incorporation, where a reversal pattern was observed. Alkali treatment enhances fiber hydrophilicity, improves compatibility with the matrix, and increases Young's modulus. Conversely, untreated fibers tend to make the material slightly more rigid [17–19]. The essential work of fracture method provides insights into impact fracture toughness in high-density polyethylene composites filled with zeolite [20], offering a means to determine fracture energy under impact conditions [20]. The synthesis of banana fiber composites involves extracting clean fibers with an efficiency of 51%, chopping them to a length of 2 mm, and treating them with NaOH to enhance integration with the polymeric matrix, improve adhesion, and elevate the degradation temperature for enhanced thermal stability. The composites that are produced have a tensile modulus of 43 GPa and a tensile strength of 891 MPa indicating that they are not easily stretched or broken [21].

Studies also show that banana fibers, when treated with alkali, improve the mechanical characteristics of polymer composites, specifically the tensile ones. A fiber-matrix weight ratio of 30–70% is ideal for composite strength, with samples subjected to alkali treatment having stronger mechanical properties than raw ones [18]. This study focuses on the need to increase the awareness of banana fibers as a reinforcement material to illustrate that it can improve composite properties for various applications [22]. However, BFRCs have been found to have better mechanical and physical characteristics. Extensive analyses of their contents and mechanical and physical properties demonstrate their suitability for numerous applications [23]. Epoxy composites reinforced with banana fiber possess higher mechanical characteristics, such as tensile, compressive, and impact strengths, with a more noticeable improvement at 20 wt.% fiber reinforcement. Furthermore, these composites also exhibit enhanced fire resistance, which cements their viability in various applications [24]. Banana fiber/glass fiber-reinforced polyester nanoclay composites exhibit enhanced mechanical properties, particularly strength, for automotive and industrial applications [25]. The transition from using agricultural waste to banana fibers in the production of fiber-reinforced composites is recommended for affordability and as an environmentally friendly approach in material production [26]. The incorporation of banana fiber reinforcement in composites demonstrates the possibility of improving the mechanical characteristics and decreasing the ecological repercussions compared to conventional materials [27]. Research carried out to determine the effects of fibers length, chemical treatments, and configuration on the properties of composite materials shows that innovation in the use of natural fibers is still ongoing [28].

Banana, along with other natural fibers, is used more frequently in composite material applications because of its renewability, biodegradability, eco-friendliness, and low cost compared to synthetic fibers [29]. Composites reinforced by banana fiber have been investigated with many matrices, and their applications are apparent in most industrial fields [31]. These studies emphasize the applicability and robustness of banana fibers in improving composite characteristics for research on the sustainable development of advanced material systems [31, 32]. It has been suggested that some coupling agents play an important role in the bonding between banana fibers and polymers, thereby increasing the efficiency of the composite and its life span [32]. The tensile properties of banana fiber-reinforced HDPE composites exhibit improvements in their physical, mechanical, and thermal properties suitable for their applications [33]. SEBS composites containing banana fibers showed improved tensile strength and stiffness when the fibers were treated with alkali, indicating that surface treatment improved composite performance [34].

The research gap on BFRCs has prioritized conducting extensive systematic reviews to map current research trends, key findings, and gaps in the field. This approach provides a comprehensive understanding of the state of the BFRC research. Developing and utilizing standardized methodologies for evaluating and comparing studies on BFRCs are crucial to ensure the consistency and reliability of findings across different research efforts. Performing meta-analyses to statistically aggregate data from multiple studies can yield more robust conclusions about the properties and performance of BFRCs, thus enhancing the overall validity of the research.

3 Methodology

3.1 Search Query

This information describes the research process for a study on innovation, especially from 2012 to 2023 [35, 36], with a focus on BFRCs, beginning with network analysis. This entails the creation of co-occurrence and co-citation networks to better understand the connections between articles and keywords. Rigorous selection criteria were devised to ensure that the chosen papers were relevant. The data from the above references were gathered and organized into an analysis-ready CSV file. RStudio is a quantitative technique for processing bibliometric data for literature reviews. The search approach used in Scopus, a massive archive of abstracts and citations, has proved critical in the research process. The infographic depicts the process of combining keywords, especially "**Banana**" AND "**Fiber**" AND "**Reinforced**" AND "**Composites**." Figure 1 illustrates the Search Query. This graphic depiction successfully depicts a methodical strategy to find and evaluate literature on the issue, exhibiting a diligent and complete research technique [37].

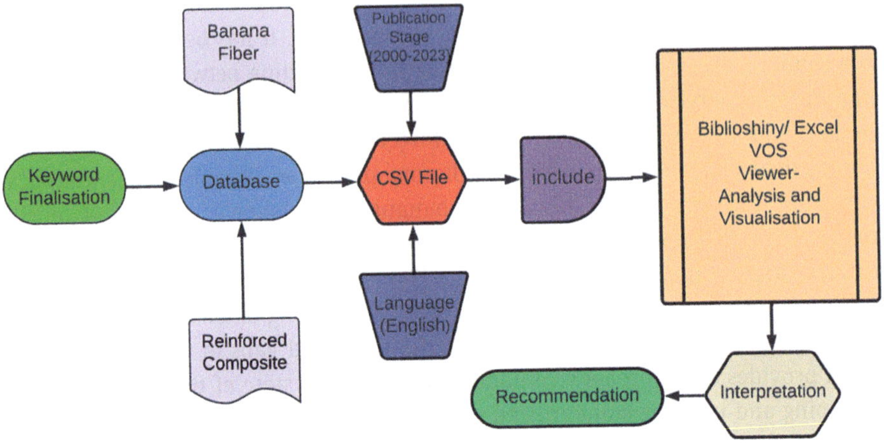

Fig. 1: Search query The TITLE-ABS-KEY (banana AND fiber AND reinforced AND composites) AND PUBYEAR > 2000 AND PUBYEAR < 2024 AND (LIMIT-TO (LANGUAGE, "English")).

3.2 Main Information

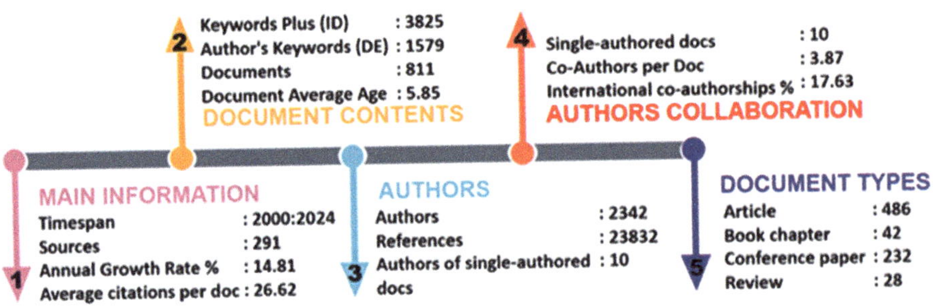

Fig. 2: Main information.

Figure 2 presents a statistical overview of the bibliometric data covering the period from 2013 to 2023. Divided into three sections – document types, author collaboration, and document content – it encapsulates key metrics. These included 1,579 author keywords, 2,342 authors, 36 single-author papers, and 3,825 keywords. The global co-authorship rate is 17.63%. Each document received an average of 26.62 citations, with 23,832 references cited. The data reflect an annual growth rate of 14.81%, encompassing 811 papers from 291 sources. While articles dominate, conference papers, reviews, and other document types were also included.

4 Bibliographic Analysis

4.1 Publications and Citation Trends

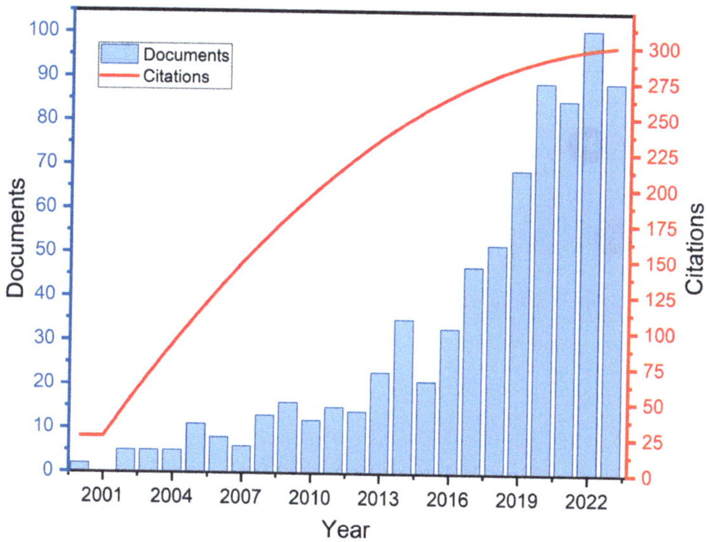

Fig. 3: Publication versus citations.

Figure 3 depicts the connection between the number of publications and citations, illustrating the annual patterns from 2001 to 2023. The blue bars, which represent the number of publications per year, exhibit a steady increase from 2001, with substantial growth beginning in 2016. This growth will reach its pinnacle around 2021–2022, with almost 100 documents published each year, reflecting the recent escalating research output. Conversely, the red line, signifying citations, demonstrates a consistent upward trajectory over the years, with a remarkable leap from 2016 onward. Citations achieve their zenith around 2022–2023, reaching approximately 300, indicating increasing recognition and impact of published research.

The association between the growing number of publications and the increasing number of citations suggests an augmenting influence of the research field. This trend signifies the expanding academic interest and scientific contributions over the past two decades, highlighting the enhanced visibility and importance of research conducted during this time period. The use of color-coding in the figure facilitates the distinction between the two variables, presenting a clear visual representation of the data.

4.2 Most Relevant Sources

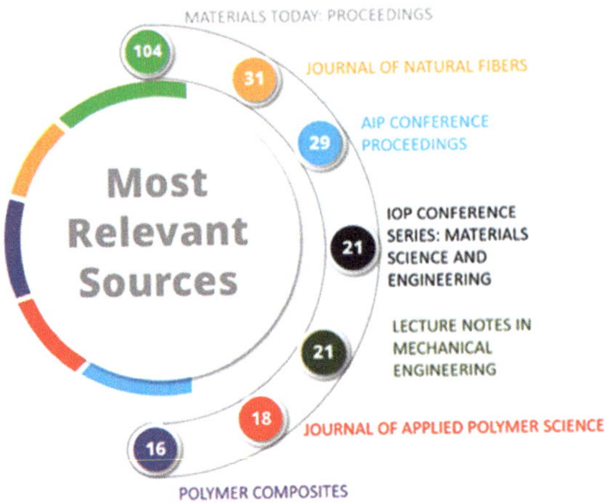

Fig. 4: Most relevant sources.

The graph in Fig. 4 displays the "**Most Relevant Sources**" for scientific publications, organized in a circular chart using different colors to represent various journals or conference series. This chart indicates the sources along with the respective number of articles and papers. **Green** signifies *Materials Today: Proceedings*, with 104 articles in the top position. **Yellow** represents the *Journal of Natural Fibers*, with 31 articles and second rank. **Blue** symbolizes the *AIP Conference Proceedings*, with 29 articles, in the third spot. **Dark Gray** indicates the *IOP Conference Series: Materials Science and Engineering*, with 21 articles, in the fourth position. **Olive green** denotes *Lecture Notes in Mechanical Engineering*, with 21 articles, in fifth position. Red represents the *Journal of Applied Polymer Science*, with 18 articles, in sixth position. **Purple** signifies *Polymer Composites*, with 16 articles, in the seventh position. The term "Most Relevant Sources" suggests the significance of these literature sections within the research field, in accordance with the overall theme. Graphical representation uses color to visually differentiate categories, facilitating rapid data interpretation.

4.3 Corresponding Author's Countries

The infographic in Fig. 5 presents bibliometric statistics on academic production in India, the USA, Malaysia, Bangladesh, and Ethiopia. India had 399 articles, of which 361 were single-country publications (SCPs) and 38 were multi-country publications (MCPs), resulting in a frequency of 0.492 and MCP ratio of 0.096. The USA followed 208 articles, 189 of which were SCPs, 19 were MCPs, with a frequency of 0.25, and an MCP ratio of 0.091. Malaysia ranked third with 39 articles, comprising 33 SCPs and 6 MCPs,

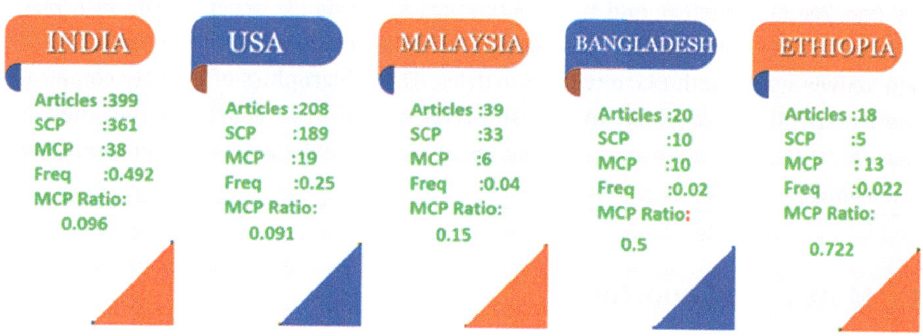

Fig. 5: Corresponding author's countries.

a frequency of 0.04, and an MCP ratio of 0.15. Bangladesh ranked fourth, with 20 articles, 10 SCPs, 10 MCPs, a frequency of 0.02, and an MCP ratio of 0.5. Ethiopia ranked fifth with 18 articles, 5 SCPs, 13 MCPs, a frequency of 0.022, and an MCP ratio of 0.722. Each country's data are presented in tabular format and clearly labeled, with an electric bulb emblem denoting concepts or research. SCP and MCP counts and ratios were compared, which may indicate research partnership trends. The frequency may represent the percentage of global publications. Color coding and style emphasize countries' contributions to and cooperation in global research, allowing for easy comparison across the board.

4.4 Most Local Cited Sources

Fig. 6: Most local cited sources.

The infographic in Fig. 6 shows the ranking of sources based on the number of citations received by the articles. Each circle in the graph corresponds to a particular source and number of articles. The top three sources were the KPR Institute of Engineering and Technology with 40 articles, followed by the Indian Institute of Technology Roorkee (34 articles), and University Putra Malaysia (32 articles). The fifth place was ANNA University and VIT University with 30 articles each, followed by Kalasalingam University in sixth place with 28 articles. The infographics effectively communicate the significant scholarly impact and relevance of these sources within their respective fields, with the use of color enhancing the visually appealing representation of the data.

4.5 Most Cited Countries

Fig. 7: Most cited countries.

Figure 7 highlights the most cited countries in the research, detailing the number of citations each country has received with color codes for clarity. **India**, represented in yellow, has 12,561 citations, demonstrating a significant research output. **Malaysia**, marked in green, has 1,548 citations, indicating its growing academic influence. **Brazil**, depicted in red, has 865 citations, indicating its substantial contribution to the research. **France**, shown in orange, followed 790 citations, underscoring its strong academic presence. **Spain**, illustrated in gray, has 291 citations, showing its impact in various research fields. **Egypt**, represented in blue, has 288 citations, reflecting active research engagement and contributions to technological advancement. These color-coded data underscore the global distribution of influential research, highlighting the leading role of these countries in advancing academic and scientific knowledge across diverse domains.

4.6 Local Impact by H-Index

Figure 8 shows the H-index scores for numerous topics or publications within the realm of materials science. *Materials Today: Proceedings* takes the top spot with an H-index of 24, followed by *Journal of Applied Polymer Science* at 16. The *Journal of Reinforced Plastics and Composites & Composites Science and Technology* ranks third and fourth, with H-index values of 11 and 10, respectively. Placing fifth in terms of the H-index is *Journal of Natural Fibers* at 10, while *Materials and Design* holds the sixth

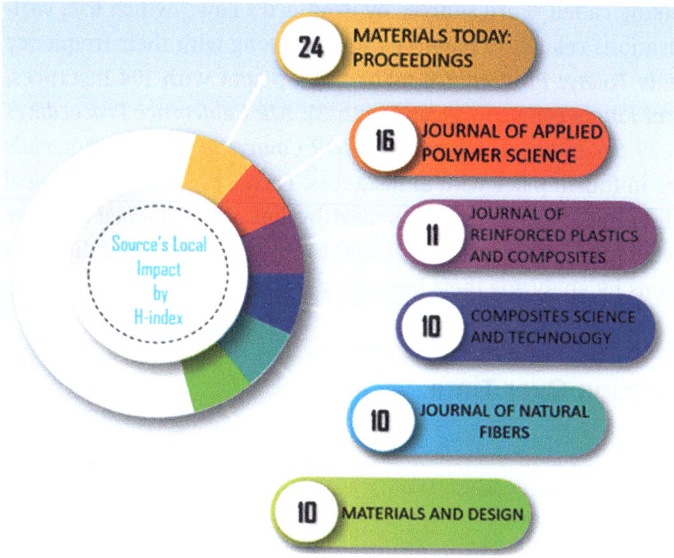

Fig. 8: Source's local impact by H-index.

spot with an H-index of 10. The rankings derived from the H-index values reflect the impact of the local community on various disciplines and journals.

4.7 Core Sources by Bradford's Law

Fig. 9: Core sources by Bradford's law.

Figure 9 displays a ranking called "Core Sources by Bradford's Law," which lists various subjects and publications related to materials science along with their frequency of occurrence. *Materials Today: Proceedings* takes the top spot with 104 instances, while *Journal of Natural Fibers* comes in second with 31. *AIP Conference Proceedings* takes third place with 29 occurrences, followed by *IOP Conference Series: Materials Science and Engineering* in fourth place with 21 instances. *Lecture Notes in Mechanical Engineering* ranks fifth with 21 occurrences, followed by *Journal of Applied Polymer Science* with 18 instances. This ranking provides insight into the influence of different locations and publications by analyzing their frequency scores.

4.8 Source's Production Over Time

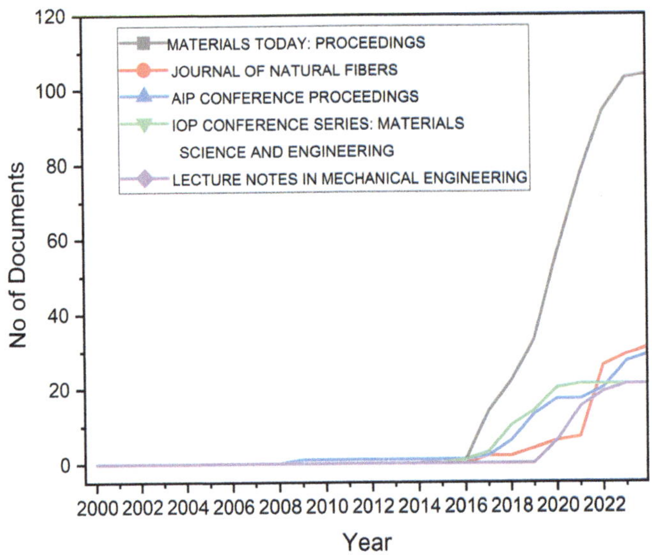

Fig. 10: Source's production over time.

Figure 10 shows the number of documents published from 2000 to 2023 in five journals/conference proceedings related to materials and engineering. Initially, publication activity was minimal until approximately 2010. A noticeable increase began around 2014–2016, reflecting growing research interest. Materials Today: Proceedings exhibit the most significant rise, with over 100 documents by 2023, indicating its growing popularity for publishing research. Other sources, such as the *Journal of Natural Fibers, AIP Conference Proceedings, IOP Conference Series: Materials Science and Engineering*, and *Lecture Notes in Mechanical Engineering*, show a more gradual increase, reaching approximately 20–40 documents each by 2023. Among these, the *Journal of Natural Fibers* and *IOP Conference Series* have slightly higher publication counts. This trend highlights a marked growth in materials and engineering research, with *Materials Today: Proceedings* standing out for their steep increase, reflecting the expanding significance of these fields in academic research.

4.9 Most Relevant Authors

Fig. 11: Most relevant authors.

Figure 11 provides a visual representation of the top four authors, ranked by their publication metrics between 2013 and 2024. The graphic shows Thomas S as the leading author, with 27 articles created, and his Articles Fractionalized Score was 7.97. The following is Sapuan S M, who is highlighted with 14 articles and an Articles Fractionalized score of 3.77 Orange. The third position is Siengchin S, represented by a blue color, who contributed 11 articles with an Articles Fractionalized score of 1.88. Lastly, in the fourth position, also represented by a red color, is Author Kumar S, who also contributed 10 articles with an Articles Fractionalized score of 2.63. This visual representation clearly conveys the significant contributions made by these authors to the research conducted during the designated timeframe. By using different colors and icons to distinguish between the authors, the graphic makes it easy to compare their contributions. The fractionalized articles provide valuable insights into the individual contributions of the authors, beyond just the number of articles.

5 Results Analysis

5.1 Most Relevant Words

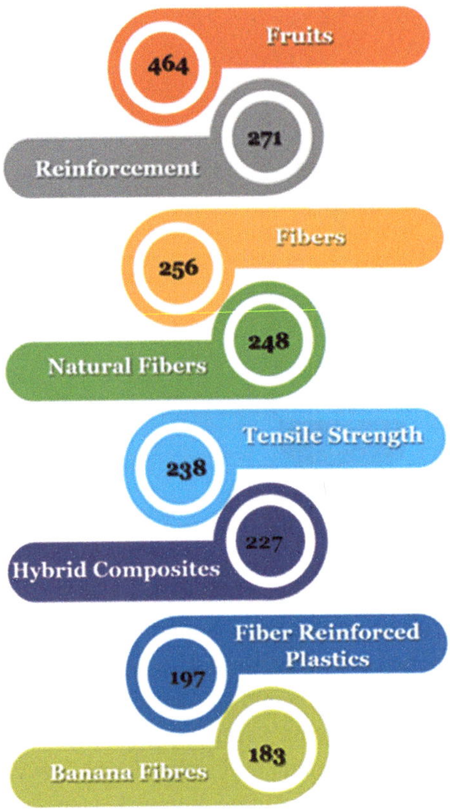

Fig. 12: Most relevant words.

Figure 12 features a series of color-coded and scaled circles that represent the prominence of specific terms within a research dataset. Each circle is proportional to the frequency of the term's appearance in the data, with the term "Fruits" standing out prominently in orange, followed closely by "Reinforcement" in gray, which is mentioned 271 times. The term "Fibers" appears 256 times in yellow, while "Natural Fibers with Green" and "Tensile Strength with Light Blue" appear 248 and 238 times, respectively, in varying shades. The arrangement of the circles emphasizes the relative significance of each term, with the most prevalent topics being highlighted in a stepped visual effect from top to bottom.

5.2 Analysis of Keywords

Tab. 1: Top 10 keywords.

S. no.	Keywords	Frequency
1.	Fruits	464
2.	Reinforcement	271
3.	Fibers	256
4.	Natural fibers	248
5.	Tensile strength	238
6.	Hybrid composites	227
7.	Fiber-reinforced plastics	197
8.	Banana fibers	183
9.	Scanning electron microscopy	169
10.	Reinforced plastics	148

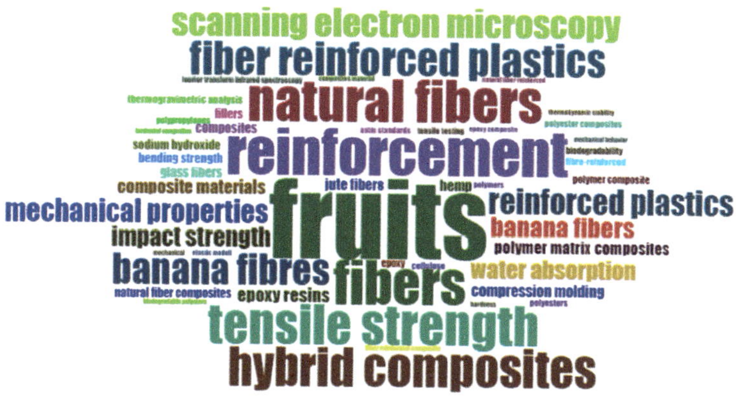

Fig. 13: Analysis of keywords.

Table 1 and Fig. 13 illustrate the strategic analysis of keywords in scholarly articles, is crucial for revealing the predominant interests and focal points of research within specific domains. The utilization of keywords in publications provides a quick reference to the core themes and highlights of research. In Fig. 10, the word dynamics within the research paper highlight the most prevalent terms. Topping the list is "**fruits**," mentioned notably 464 times, closely followed by "**reinforcement**" with 271 mentions. Another significant term is "**fibers**," which appear prominently at a frequency of 256. Additionally, terms such as "**natural fibers**" and "**tensile strength**" carry importance, with frequencies of 248 and 238, respectively. These keywords underscore the emphasis of research on technological advancements and material sciences, shaping the future of automotive applications using 3D printing.

5.3 Bibliographic Coupling of Countries

Tab. 2: Top countries.

Country	Documents	Citations	Total link strength
India	556	16,306	6,103
Saudi Arabia	22	485	2,922
Malaysia	59	1,979	2,135
Thailand	16	477	1,775
Ethiopia	24	180	1,315
United Kingdom	13	364	1124
Egypt	15	651	996
Australia	9	630	905
USA	20	514	891
Bangladesh	26	327	880

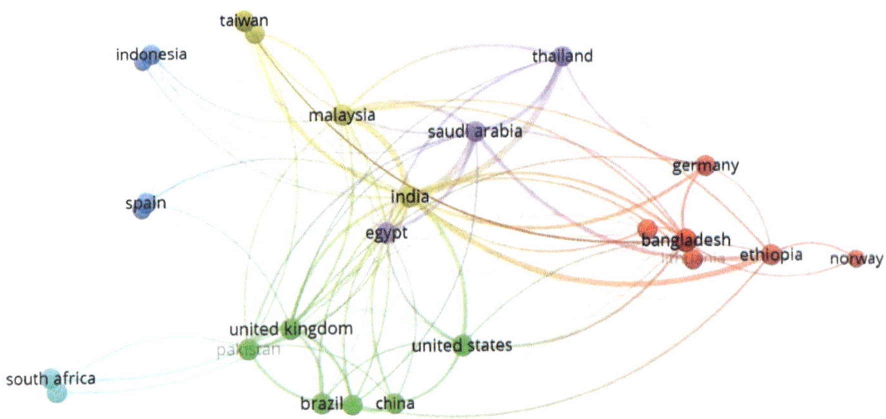

Fig. 14: Bibliographic coupling of countries.

Table 2 and Fig. 14 depict the "Bibliographic Coupling of Countries" as identified by the authors in their study. The authors established specific criteria for selecting journals, requiring a minimum of 10 documents and 10 citations from each country. Of the 81 countries analyzed, only 25 met the inclusion criteria. This study employed quantitative network indicators to characterize the relationships between clusters of journals with related content. VOS viewer software was used to analyze the Bibliographic Coupling of Countries' data derived from the article. The resulting analysis categorized the articles into six distinct clusters, as shown in Fig. 14, with different colors. India leads the clusters with 556 documents, 16,306 citations, and a total link strength (TLS) score of 6,103. Saudi Arabia closely followed 22 documents, 4,850 citations, and a TLS score of 2,922. Malaysia ranks third, with 59 documents, 1,979 citations, and a TLS score of 2,135. Thailand ranks fourth, with 16 documents, 477 citations, and a TLS score of 1,775. Ethiopia, the United Kingdom, Egypt, Australia, the USA, and Bangladesh were ranked fifth to tenth, with TLS scores of 1,315, 1,124, 996, 905, 891, and 880, respectively.

5.4 Authors' Keywords of Co-occurrence

Tab. 3: Authors' keywords.

Keywords	Occurrences	Total link strength
Mechanical properties	146	185
Banana fiber	158	145
Composites	66	74
Hybrid composites	47	63
Natural fiber	60	51
Water absorption	29	50
Natural fibers	59	49
Epoxy	40	48
SEM	28	43
Banana fiber	49	41

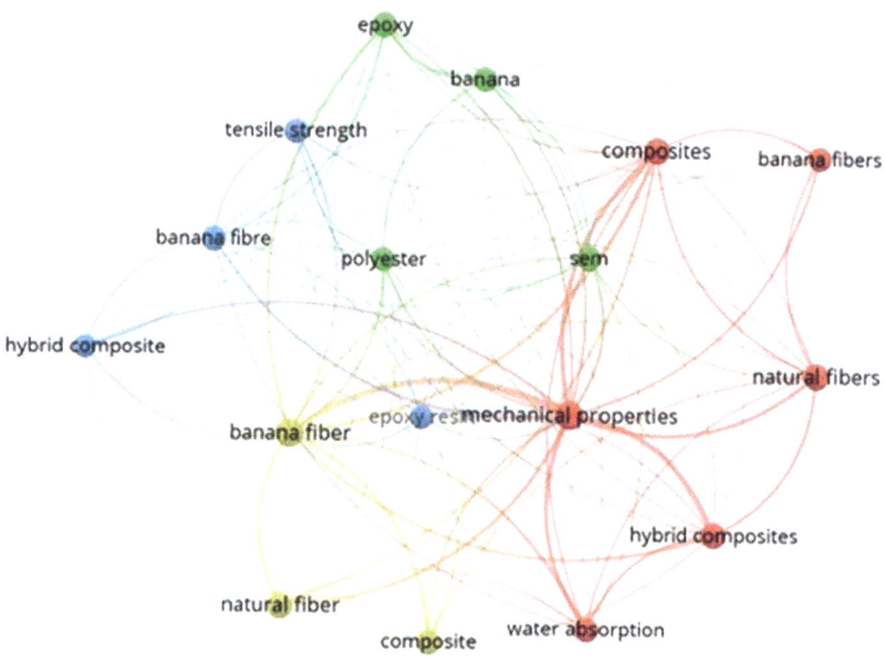

Fig. 15: Authors' keywords.

Figure 15 and Tab. 3 illustrate the "Authors' Keywords of Co-occurrence" identified by the researchers in their study. The keywords selected for the analysis were those with a minimum of 20 co-occurrences. Out of a total of 1,584 terms, only 17 met the specified criteria.

The size of each node is proportional to its rate of occurrence. The connections between the nodes signify the co-occurrence of these keywords within the same context. Leading the clusters is "Mechanical properties" with 146 occurrences and a TLS of 185. Following closely is "Banana fiber," with 158 occurrences and a TLS of 145. "Composites" ranks third with 66 occurrences and a TLS of 74, while "Hybrid composites" ranks fourth with 47 occurrences and a TLS of 63. Additional keywords such as "natural fiber," "water absorption," "natural fibers," "epoxy," "sem," and "banana fibre" are ranked fifth to tenth, with TLS scores of 51, 50, 49, 48, 43, and 41, respectively.

6 Summary and Conclusions

This study aims to conduct an extensive investigation of the trends in BFRCs using scientometric analysis. By analyzing publishing, annual research production, and citation patterns, this study aims to reveal significant trends in countries, publications, authors, sources, and studies that have significantly contributed to this research area. The following conclusions drawn from this analysis:

- There has been a significant increase in research on banana fibers in reinforced composite materials since its inception in 2016.
- India leads the way in research output, followed by the USA and Malaysia. However, these three countries combined contributed to over 80% of the publications, highlighting their potential dominance. India also leads in terms of citations received, with China following closely.
- Key journals publishing research in this area include *Materials Today: Proceedings*, the *Journal of Natural Fibers*, and *AIP Conference Proceedings*.
- Scientometric analysis reveals core interests such as banana fibers, composite materials, and polymer composites. Additionally, themes like "hybrid composites," "natural fiber," and "nanocomposites" are gaining or losing traction.
- Areas such as reinforcement, composite materials, and natural fibers hold significant promise for future advancements based on the identified specialty themes.

7 Limitations and Future Research Directions

7.1 Limitations

The present study utilized Scopus as its primary source database, although several other databases, such as Web of Science and Lens, are also available. The three databases, org, Dimensions, and PubMed, each employed different indexing methods, policies, and query interface execution systems. It is crucial to select an appropriate database because it plays a significant role in determining study outcomes. The search was conducted in July 2024, and it should be noted that the databases were periodically reviewed, which may result in changes to the results of subsequent searches. In light of the limited research on the potential of artificial intelligence in harnessing sustainable materials, future studies may consider employing multiple databases over an extended period and analyzing a larger number of sources. Additionally, scientometric analysis can be used to address this gap in our understanding.

7.2 Future Research Directions

Future research on BFRCs should focus on integrating artificial intelligence for predictive modeling and optimization, utilizing multiple databases over extended periods to gain a comprehensive understanding of the field, and exploring hybrid composites that combine banana fibers with other materials to enhance properties and applications. Detailed sustainability assessments and life cycle analyses should be conducted to compare the environmental impacts of BFRCs with those of conventional materials, while extensive mechanical and durability testing are needed to assess their performance under various conditions. Research should also investigate the feasibility of scaling-up production for industrial applications, conduct policy and market analyses to address adoption barriers, and explore bio-functionalization and chemical treatments to improve fiber-matrix bonding.

References

[1] Yu F. Study on Preparation and Application of Banana Fiber-Based Composites. J Phys Conf Ser. 2023;2539:012093. doi: 10.1088/1742-6596/2539/1/012093.

[2] Benson MJ. Characterization of False Banana Fiber as a Potential Reinforcement Material for Geopolymer Composites. Green Energy Technol. 2023;49–63. doi: 10.1007/978-3-031-33610-2_3.

[3] Islam N, Gafur MA. Matrix-Material Fabrication Technique and Thermogravimetric Analysis of Banana Fiber Reinforced Polypropylene Composites. J Build Mater Sci. 2023. doi: 10.30564/jbms.v5i2.5700.

[4] Pal A, Kumar ND, Gautam N. Banana Fiber Reinforced Concrete. Int J Sci Technol Eng. 2023;11(6):4650–71. doi: 10.22214/ijraset.2023.54493.

[5] Ramu S, Senthilkumar N, Rajendran S. MWCNT Filled Banana-Rice Husk Epoxy Hybrid Natural Fiber Polymer Composites. Mater Sci Forum. 2023;1082:53–63. doi: 10.4028/p-8m528f.
[6] Kelly-Walley J, Ortega Z, McCourt MP, Millar B, Suárez L, Martin P. Mechanical Performance of Rotationally Molded Multilayer mLDPE/Banana-Fiber Composites. Materials (Basel). 2023;16 (20):6749. doi: 10.3390/ma16206749.
[7] Bekraoui N, El Qoubaa Z, Chouiyakh H, Faqir M, Essadiqi E. Banana Fiber Extraction and Surface Characterization of Hybrid Banana Reinforced Composite. J Nat Fibers. 2022;19(16):12982–95. doi: 10.1080/15440478.2022.2080789.
[8] James J, Busfield C. Study on Mechanical Properties of Banana Fiber-Reinforced Poly (Lactic Acid) Composites. Int J Chem Eng. 2022;2022:1–7. doi: 10.1155/2022/8485038.
[9] Anonymous. Synthesis and Characterization of Mechanical and Physical Properties of False Banana (ʼənäsätə) Fiber Reinforced Composite Material. 2023;1(1):49–61. doi: 10.59122/134cfc10.
[10] Singh S, Naik N, Sooriyaperakasam N, Iyer TSR, Agarwal C, Tirupathi J, et al. Banana Fiber Reinforced Polymeric Hybrid Composites: An Overview. 2022;1(2):36–49. doi: 10.57159/gadl.jcmm.1.2.22011.
[11] Piyalak N, Leekrajang M. Extraction, Improvement of Banana Stem and Water Hyacinth Cellulose Fibers as Reinforcement in Cementitious Composites. J Met Mater Miner. 2022;32.
[12] Aarjoo S, Neena C, Prince S. Banana Fibre: A Natural and Sustainable Bioresource for Eco-friendly Applications. 2021.
[13] Naveen R, Ajay MB, Ajay Kk. Mechanical Characterization of Unidirectional Banana–Glass Fiber-Reinforced Hybrid Composites. n.d.
[14] KS, RA, SA, Ragul VR. Mechanical Behaviour of Natural Coconut/banana Fiber Reinforced Composites. n.d.
[15] José AB, De Barros RTP, Cândido AOT, Dos Reis D, Martins CR. The Incorporation of Untreated and Alkali-treated Banana Fiber in SEBS Composites. 2020.
[16] Aarjoo S, Neena C, Prince S. Banana Fibre: A Natural and Sustainable Bioresource for Eco-friendly Applications. 2021.
[17] Subba R, Gvr, Suresh GM. Mechanical Properties of Banana Fibre Reinforced Polymer Composites. n.d.
[18] Putu HS, Sulistyaningsih D. Impact-specific Essential Fracture Work of Banana Fiber Reinforced Low-density Polyethylene Composites. IMSF. 961. n.d.
[19] Brkić U, Miličević M, Pineda R, Bordón P. Thermal and Mechanical Characterization of Banana Fiber Reinforced Composites for Its Application in Injection Molding. Materials (Basel). 2020;13.
[20] Gohil TG, Sutar VS. Applicability of Banana Fiber as Reinforcement in Composites. n.d.
[21] Ramesh M. Banana Fibers, Their Composites and Applications. IPF; n.d.
[22] Tuan A, Thi HN. Banana Fiber-reinforced Epoxy Composites: Mechanical Properties and Fire Retardancy. 2021.
[23] R R, B K, Madhubalan R. Study the Mechanical Behaviour of Banana Fiber/glass Fiber Reinforced Polyester Nano Clay Composites. AIP Conf Proc. n.d.
[24] Faheem A, Farooq A, Salma S, Yasir N, Ahmad S. An Economical and Environmentally Benign Approach to Extract Banana Fibres from Agricultural Waste for Fibre Reinforced Composites. 2022.
[25] Pawar C, Bhat K, Goud J. Review on Banana Fibre-reinforced Composites. 2021.
[26] Raphael S, Onukwuli DO, Ernest ME, Mathew CM, Agu PC. The Exchange of Musa Spp. Fibre in Composite Fabrication: A Systematic Review. Bull Natl Res Centre. 2021;45.
[27] Shashank KS, Anil KJ, Vignesh P, Nilesh S, Chinnaiyan P. Investigations on Tensile and Flexural Characteristics of Fly Ash and Banana Fiber-reinforced Epoxy Matrix Composites. Eng Trans. 2020;68.
[28] M S, Mashrafi BM, Farzana KR, Sazia S, Monika M, Samina A. Fabrication and Characterization of Banana Fiber Reinforced Unsaturated Polyester Resin Based Composites. 2020.

[29] Raza A, Khan MY, Iqbal R, Ahmad U, Koc M. Utilization of Banana Fiber-reinforced Hybrid Composites in the Sports Industry. Materials (Basel). 2020;13.

[30] Kumar RK, Venkatesh NR, Agarwal PK Characterization of banana fiber-reinforced polypropylene composites. 2020.

[31] Hossain R, Ferdous K, Gafur MA, Hossain MA, Ahmed F. Study of the Physical, Mechanical and thermal Properties of Banana Fiber Reinforced Hybrid Composites. Mater Sci Appl. 2020;11.

[32] José AB, Rtp DB, Cândido AOT, Dos Reis D, Martins CR. The Incorporation of Untreated and Alkali-treated Banana Fiber in SEBS Composites. Polímeros. 2020;30.

[33] Baas J, Schotten M, Plume A, Côté G, Karimi R. Scopus as a Curated, High-quality Bibliometric Data Source for Academic Research in Quantitative Science Studies. Quant Sci Stud. 2020;1:377–86. doi: 10.1162/qss_a_00019.

[34] Ravindran G, Mahesh V, Bheel N, Chittimalla S, Srihitha K, Sushmasree A. Usage of Natural Fibre Composites for Sustainable Material Development: Global Research Productivity Analysis. Buildings. 2023;13:1260. doi: 10.3390/buildings13051260.

[35] Srivastav P, Vutukuru M, Ravindran G, Awad MM. Biofortification – Present Scenario, Possibilities and Challenges: A Scientometric Approach. Sustainability. 2022;14:11632. doi: 10.3390/su141811632.

[36] Kumar A, Kumar P, Sebaey TA, Gambhir V, editors. Tools, Techniques, and Advancements in Engineering Materials Science. IGI Global; 2024 Dec 24.

[37] Kumar A, Rajak DK, Kumar P, Kumar A, editors. Advances in Sustainable Biomaterials: Bioprocessing 4.0, Characterizations, and Applications. CRC Press; 2024 Oct 28.

[38] Patel S, John D, Kollarath RM, Kumar P. Sustainable materials for urban streets: Trends, challenges, and case studies. In: Advances in Sustainable Materials. Elsevier; 2025 Jan 1. pp. 515–30.

[39] Avani S, Sharma I, Nagpal S, Saini P, Kumar P, Kumar A. Applications of Biopolymers in the Agriculture and Food Industry. In: Advances in Sustainable Biomaterials. CRC Press; pp. 323–54.

[40] Geeri S, Kolakoti A, Kumar P, Gudimetla A. Experimental and optimization analysis on mechanical properties of polymer-based natural composites. In: Advances in Sustainable Materials. Elsevier; 2025 Jan 1. pp. 229–53.

[41] Yadav R, Praksah C, Kumar P, Kumar A. Identifying factors affecting the choice of sustainable materials: A factor analysis approach for sustainable development. In: Advances in Sustainable Materials. Elsevier; 2025 Jan 1. pp. 289–304.

[42] Koushiki C, Christine J, Kumar P. Sustainable materials as a novel source for biomedical applications. In: Advances in Sustainable Materials. Elsevier; 2025 Jan 1. pp. 471–85.

Ruchira Srivastava*, Ayushi Thakur, Parveen Kumar, and Ajay Kumar

Exploring the Applications of Smart Materials in the Medical and Healthcare Sectors

Abstract: The development of SMART (self-healing, multifunctional, adaptive, responsive, and tunable) materials has been extremely beneficial for biomedical and healthcare applications. Such materials have unique properties that show sensitivity to changes in their environment, making them quite attractive for many different types of medicinal uses. This study investigates the advancements and challenges associated with smart materials in the biomedical domain. Significant progress has been made in the creation, synthesis, and analysis of intelligent materials intended for use in biomedicine and healthcare wearables over the last several years. Taking cues from natural regeneration mechanisms, self-healing materials have been used to build medication delivery systems, implants, and wound-healing scaffolds. The continuous progress in smart material science presents enormous potential to revolutionize biological applications. It is anticipated that the development of biomaterials, bioengineering, and nanotechnology would be crucial to the creation of materials with improved attributes and capacities. Precision medicine and personalized healthcare may see new possibilities if smart materials are combined with cutting-edge technologies like gene editing, 3D printing, and microfluidics. To overcome the current challenges, researchers, doctors, regulatory bodies, and industry partners must work together to successfully translate smart materials from the lab to the clinic.

Keywords: Smart materials, biomedical applications, healthcare wearable, 3D printing

1 Introduction

The notion of materials, systems, and structures that are "smart" or "intelligent" has been around for a while. The development of structures that actively and continuously evaluate their performance and themselves has advanced significantly in recent years, simulating biological systems with their integrated designs and adaptive capa-

*Corresponding author: Ruchira Srivastava, Department of Mechanical Engineering, Amity University, Greater Noida 201310, Uttar Pradesh, India, e-mail: engg.ruchira09@gmail.com
Ayushi Thakur, Department of Mechanical Engineering, Amity University, Greater Noida 201310, Uttar Pradesh, India
Parveen Kumar, Skill Department of Automotive Studies, Shri Vishwakarma Skill University, Dudhola, Palwal 121102, Haryana, India
Ajay Kumar, Department of Mechanical Engineering, School of Core Engineering, Faculty of Science, Technology and Architecture, Manipal University Jaipur, Rajasthan, India

https://doi.org/10.1515/9783111515717-011

bilities [1]. The trans-disciplinary and interdisciplinary field of smart materials encompasses a range of enabling technologies, including materials, control, information processing, sensing, actuation, and damping, as well as system integration across a wide range of industrial applications. Fiber-optic sensor systems, magnetostrictive materials, shape memory alloys (SMA) and polymers, micro-machined electromechanical systems (MEMS), conductive polymers, and electrostrictive and magnetostrictive ceramics are just a few of the many technologies that fall under the broad heading of smart materials and structures, and each is at a different stage of commercialization [2]. It is anticipated that chromogenic materials and systems, electrorheological and magnetorheological fluids, and biometric polymers and gels will be more widely used in the upcoming wave of smart technologies [52–55].

A class of materials known as smart materials has emerged in the realm of biomedical applications as a consequence of the fusion of cutting-edge materials research with the requirements of healthcare. The word "smart" stands for self-healing, multi-functionality, and adaptability, which are the qualities of these materials adapt to shifting circumstances, take in stimuli, and change in accordance with needs [3]. Smart materials offer a variety of features and capabilities that demonstrate dynamic reactions to changes in their surroundings, demonstrating a noteworthy combination of breakthroughs and necessity in the area of healthcare. The dynamic reaction shown in this situation has opened up exciting possibilities for changing how we approach diagnosis, treatment, and medical supervision [56–59]. The concept of such intelligent materials has its roots in the larger spectrum of materials science, where researchers have long attempted to create materials with unique properties to meet specific requirements. However, there are several challenges and barriers associated with using smart materials in the biological area [4, 60–64]. The development of materials that can adapt, react, and integrate with the complex biological systems of the human body is necessary due to the increasing complexity of healthcare requirements. However, even with all of its promise, there are still obstacles in the way of smart materials' transfer from laboratory experiments to practical medical uses [5]. Some of the characteristics that show promise in clinical settings are the integration of multi-functionality, the stability of self-healing processes, and the numerous challenges that need to be handled for these materials to fully integrate biocompatibility. Additionally, to ensure that smart materials are widely used in patient care, their scalability and affordability are crucial factors [6].

When we look at the complex world of smart materials in the context of healthcare applications, as shown in Fig. 1, we can see that the nexus of materials engineering, biology, engineering, and medicine offers unparalleled potential for healthcare development. Comprehending the achievements and failures will enable us to meticulously strategize the utilization of smart materials' potential to transform healthcare in the future [7]. It is feasible to overcome the challenges and realize the promise of smart materials by cooperative efforts, creative solutions, and thorough study, therefore establishing a medical setting that is more flexible, individualized, and effective. Smart (self-healing, multifunctional, adaptive, responsive, and tunable) materials are being researched as promising new solutions to challenging healthcare problems as a result of the convergence of biomedical engineering and materials science [8]. The aforementioned convergence is fueled by the growing demand for materials with the capacity to diligently interact with ecological systems, adjust to changing ecological conditions, and provide tailored convenience for a wide range of therapeutic uses. Smart materials have the potential to revolutionize patient care, diagnosis, and treatment in the biomedical field, bringing about a new era of precise and personalized medicine.

Engineering, chemistry, biology, materials science, and other scientific disciplines all contribute to the study of smart materials [9]. Through the effective use of these Trans disciplinary techniques, researchers have created materials that display characteristics similar to the dynamic behavior of living things. This discovery has the prospective to completely alter clinical practices by giving rise to materials that can interact, heal, and adapt to organism's tissues in ways that were formerly unthinkable [10].

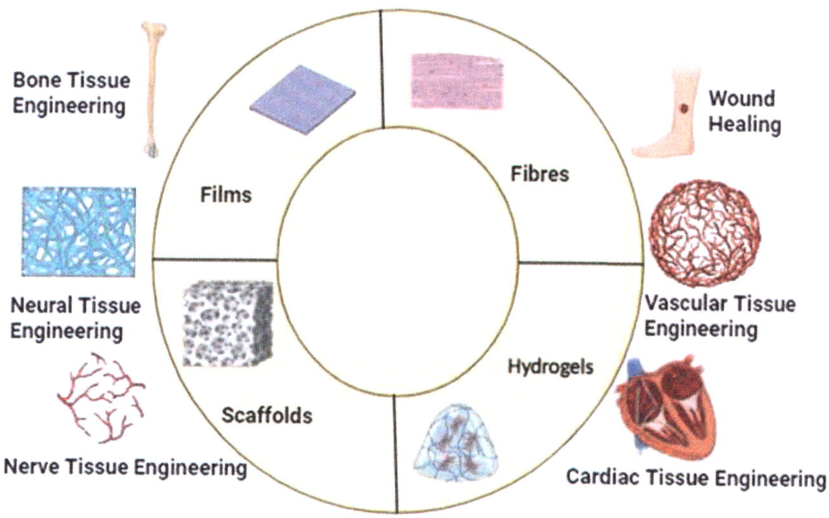

Fig. 1: Realm of SMART materials.

A broad family of materials known as "intelligent materials" has special qualities that allow them to react to environmental changes intelligently and adaptively. The intrinsic capacity of smart materials to autonomously go through repair procedures or restore their structural integrity is inherent. The possibility of autonomic-healing materials has significant applications in the domains of medication delivery, wound dressing, and implants. These materials are inspired by the principles of regeneration seen in nature. The longevity and functional capabilities of biomedical equipment might be significantly improved by being able to fix small faults at the micro-scale level. The capacity of multifunctional materials to include several functions into a single structure – often including monitoring, therapeutic, and diagnostic capabilities – defines them. Utilizing multifunctional smart materials to consolidate various features enables more targeted and effective medical treatments, which, in turn, improve treatment outcomes and enhance patient experiences [11]. The market for wearable and implantable medical devices is growing as a result of the aging population and the emergence of chronic degenerative diseases. These devices have a wide range of biological applications, including monitoring and diagnostics, pharmaceutical delivery and therapy, surgery, and rehabilitation. Tracking the progression of an illness and subsequently delivering the required treatments can be made easier with ongoing human activity monitoring. As information technology advances, this ecosystem is impacting smart healthcare and facilitating patient care both within and outside of hospitals [12]. Depending on the device application, a range of rigid and flexible materials has been employed to date to build these devices, meeting certain requirements such as biocompatibility, sensitivity, selectivity, form factor, stretchability, durability, washability, and self-healing properties. At the absolute least, medical devices contain some sort of transducer (sensor or actuator) that interacts with biological substances like blood, tears, or sweat or comes into direct touch with skin or organs [13]. The aging population and rise in chronic illnesses make people more aware of their health, and patients turn into "health consumers" who seek out better ways to manage their conditions. People's perceptions of health services are changing from traditional hospital-centered ones to patient-centered ones, and this is driving the advancement of telemedicine research from traditional e-health to m-health and, ultimately, ubiquitous healthcare [14]. Currently, the most promising solutions for increased dependability and lifespan efficiency involve the use of intelligent structures and materials. Understanding and controlling the composition and microstructure of any novel material is the ultimate goal of this field of study since it is crucial to the development of superior smart materials. The knowledge gained from gathering information on how a material's crystal core structure behaves under different heating and cooling circumstances will speed up the development of new materials with a greater range of applications [15].

2 Innovations in the Realm of Smart Materials for Medical and Healthcare Applications

Medical and healthcare procedures have greatly improved as a result of the creation of smart materials. These materials, designed to interact dynamically with their environment, offer innovative solutions that improve diagnosis, treatment, and patient care [16]. In Tab. 1, we find the latest advancements and their impacts in various medical and healthcare applications. A paradigm shift has occurred due to the development of manifold materials in the realm of healthcare practices, which offers previously unheard-of opportunities to revolutionize treatments and diagnostics. A noteworthy innovation is the amalgamation of medicinal and diagnostic capabilities into a single material substrate [17]. The long-standing problem of successfully integrating various diagnostic and therapeutic approaches, which usually need different procedures and resources, is what this integration attempts to address. The incorporation of these features in multifunctional materials not only expedites medical processes but also has the potential to improve treatment outcomes by enabling real-time monitoring and adaptive treatments [18].

Tab. 1: Latest advancements in smart materials.

Materials	Types of materials	Advancement	Applications	Examples
Self-healing materials	Self-healing polymers	Development of polymers that can autonomously heal small cracks and damages.	Used in medical implants, these materials can repair themselves, thus extending the life and reliability of devices like orthopedic implants and cardiovascular stents.	Polyurethane-based self-healing materials that respond to physiological temperatures.
Multifunctional materials	Multifunctional nanoparticles	Creation of nanoparticles capable of simultaneous imaging, therapy, and sensing.	In cancer treatment, these nanoparticles can deliver drugs, provide real-time imaging, and monitor the therapeutic response.	Gold nanoparticles conjugated with drugs and fluorescent markers for targeted cancer therapy and imaging.
	Multifunctional wound dressings	Development of dressings that combine antimicrobial activity, moisture control, and drug delivery.	Advanced wound care for chronic wounds, such as diabetic ulcers.	Hydrogel dressings that release antibiotics and growth factors in response to wound conditions.
Adaptive materials	Adaptive hydrogels	Hydrogels that alter their characteristics in reaction to external stimuli like temperature, pH, or glucose concentrations.	These hydrogels, which are utilized in insulin delivery systems, have the ability to release insulin in reaction to blood glucose levels.	pH-sensitive hydrogels that release medications into the acidic tumor environment for targeted drug delivery in cancer treatment.

Tab. 1 (continued)

Materials	Types of materials	Advancement	Applications	Examples
Responsive materials	Thermo-responsive polymers	Polymers that undergo a phase change in response to temperature variations.	Drug delivery systems that release medication when a certain temperature threshold is reached.	*N*-isopropylacrylamide (NIPAAm) polymers are used in controlled drug release for localized hyperthermia treatments.
	pH-responsive materials	Substances that react to pH variations by changing their characteristics.	Targeted drug delivery in the gastrointestinal tract, where the pH varies significantly.	pH-sensitive polymer coatings for oral drug delivery that protect the drug from stomach acid and release it in the intestines.
Tunable materials	Tunable hydrogels	Hydrogels with adjustable mechanical properties to match the surrounding tissue.	Regenerative medicine and tissue engineering, which produce scaffolds that replicate the mechanical characteristics of natural tissues.	Hydrogels with tunable stiffness for use in cartilage and bone regeneration.
	Adjustable drug delivery systems	Systems that allow fine-tuning of drug release rates and dosages.	Personalized medicine, enabling customized treatment plans based on patient-specific needs.	Implantable drug delivery devices with adjustable release mechanisms, controlled by external stimuli such as magnetic fields.
Wearable and implantable devices	Flexible electronics	Development of biocompatible, flexible electronics that can conform to the body.	Vital indicators like blood pressure, temperature, and heart rate are continuously monitored.	Flexible sensors are integrated into wearable health monitors and smart bandages.
	Implantable biosensors	Creation of biosensors that provide real-time monitoring of biochemical markers.	Monitoring glucose levels in diabetic patients or detecting early signs of infection.	Implantable glucose sensors that provide continuous glucose monitoring and insulin delivery.

Tab. 1 (continued)

Materials	Types of materials	Advancement	Applications	Examples
Biomimetic materials	Synthetic biomimetic materials	Materials that mimic the properties and functions of natural tissues.	Used in regenerative medicine and prosthetics to create more natural and functional replacements.	Synthetic tendons and ligaments are made from biomimetic polymers that replicate the elasticity and strength of natural tissues.

This development depends on multifunctional materials' simultaneous function as drug transporters, sensors, and imaging agents. These substances can identify and measure certain disease biomarkers or physiological alterations with a high degree of accuracy and sensitivity in the diagnostics sector [19]. Therapeutic efforts can then be informed by the diagnostic data, ensuring that each patient's unique needs are met by customized therapy. As an illustration, a flexible substance may precisely identify the location and dimensions of a tumor while also delivering specific therapeutic chemicals just to the affected area. By using this targeted approach, the goal is to minimize damage to healthy tissues and maximize the effectiveness of treatment for diseased regions [20]. Applications for smart materials range from the fashion, aerospace, and building sectors to the electronics and medical sectors. These materials, especially soft functional smart materials, are responsive to outside stimuli, which make them appropriate for compliant sensing and inventive programmable motion. Numerous studies have been conducted on soft robotics, haptic electronics, biomedical devices, and smart textiles using soft sensors and actuators made from these materials. However, the high density of a highly interconnected network of transducers (such as sensors and actuators) and the need to charge them with big, heavy batteries sparked the idea of energy-autonomous systems. To extend the battery life of these gadgets, harvesting ambient energy from the surroundings is a feasible alternative [21]. This could be accomplished by developing and integrating energy harvesting and storage devices with transducers. In the context of smart healthcare, IoT and self-powered flexible systems collaborate to enhance illness management, accuracy, cost-effectiveness, and patient experience. The Internet of Things, biotechnology, and artificial intelligence are essential elements of smart healthcare [22].

Biotechnology-based wearable and implantable devices provide effective solutions for smart healthcare in both therapeutic and diagnostic domains. The increasing development of biomedical equipment, including neurostimulators, brain-machine interfaces, and smart controlled systems, has expanded the scope of smart healthcare applications. With the advent of smart healthcare, well-developed ideas and frameworks have emerged. There is still a lot of space for improvement, though, as new issues and technologies arise [23]. A lot of new difficulties are also appearing. There are currently no programmatic papers or macro guidelines in smart healthcare, which results in hazy development objectives and eventually resource waste. Furthermore, data integrity has to be improved at medical institutions because there are inconsistent standards between companies and countries. Compatibility issues also exist between various systems and gadgets [24]. From the standpoint of the patient, there are concerns related to privacy breaches and personal information, and smart healthcare is not adequately regulated by applicable laws. Even some users find it challenging to operate the devices. Strictly speaking, a few smart healthcare-related technologies are still in the experimental phase and need substantial financing to be updated and maintained. If used carelessly, there's an additional unidentified risk [25].

The advancements in smart materials have opened new frontiers in medical and healthcare applications, offering enhanced functionality, responsiveness, and adaptability. These materials are transforming traditional medical treatments, enabling more effective therapies, improving patient outcomes, and paving the way for personalized medicine [26]. However, challenges such as ensuring long-term biocompatibility, obtaining regulatory approvals, and achieving large-scale manufacturing need to be addressed to fully realize their potential. Continued research and innovation in this field are essential to overcome these obstacles and further integrate smart materials into clinical practice [27]. Additionally, a level of personalization made possible by the use of multifunctional materials was previously unattainable with traditional treatment procedures. These materials allow for the dynamic alteration of treatment regimens by continuously evaluating treatment responses through the use of diagnostic capabilities. The ability to adjust is crucial in the treatment of cancer since individual patients differ greatly in the characteristics of their tumors and how they respond to various therapies. With real-time feedback, multifunctional materials can optimize therapy dosages, timing, and combinations, potentially increasing therapeutic efficacy and reducing side effects [28].

3 Challenges in the Realm of Smart Materials for Medical and Healthcare Applications

The development and integration of smart materials in medical and healthcare applications offer significant potential benefits. However, several challenges need to be addressed to fully realize their potential. These challenges span across biocompatibility, long-term stability, scalability, regulatory hurdles, and cost, among others [29]. Some of the areas of challenges as faced by smart material with issues in detail and solutions are summarized in Table 2.

Tab. 2: Challenges with solutions as faced by smart materials in medical and healthcare sectors.

Area of challenge	Issue	Details	Solutions
Biocompatibility	Ensuring that SMART materials are non-toxic and do not provoke an adverse immune response when introduced into the human body.	Many SMART materials, particularly novel synthetic polymers and nanoparticles, may cause unforeseen biological reactions. Comprehensive biocompatibility testing is essential to prevent inflammatory responses, toxicity, and long-term health issues.	Extensive in vitro and in vivo testing, surface modifications to reduce immune response, and the use of naturally derived materials that is more likely to be biocompatible.
Long-term stability and durability	Maintaining the functionality and structural integrity of SMART materials over extended periods within the body.	Materials may degrade, lose their responsiveness, or fail to self-heal after prolonged exposure to a biological environment.	Development of more robust materials with enhanced durability, long-term biocompatibility studies, and the incorporation of stabilizing agents to preserve material properties.
Scalability and manufacturing	Scaling up the production of SMART materials while maintaining consistency, quality, and performance.	Many SMART materials are complex and involve sophisticated manufacturing processes that are difficult to replicate on a large scale.	Advances in manufacturing technologies, such as 3D printing, automated production processes, and collaboration with industrial partners to scale up production efficiently.
Regulatory approval	Navigating the complex regulatory landscape to achieve approval for clinical use.	To guarantee safety and efficacy, regulatory agencies like the FDA need thorough testing, which can be expensive and time-consuming.	Early engagement with regulatory agencies, thorough documentation of safety and efficacy, and investment in regulatory expertise to streamline the approval process.
Cost and accessibility	High expenses related to the creation and manufacturing of sophisticated SMART materials.	The sophisticated nature of these materials often leads to higher costs, which can limit their accessibility and widespread adoption in healthcare.	Research into cost-effective production methods, government and private funding to support development, and efforts to reduce material costs through improved efficiency and economies of scale.

Tab. 2 (continued)

Area of challenge	**Issue**	**Details**	**Solutions**
Integration with existing medical systems	Integrating SMART materials with current medical devices and systems.	Ensuring compatibility and seamless integration with existing medical infrastructure and practices can be challenging.	Collaborative development with medical device manufacturers, ensuring interoperability standards, and providing training for healthcare professionals on the use of new materials and devices.
Ethical and social considerations	Addressing ethical and social implications related to the use of SMART materials in healthcare.	Concerns about long-term impacts, equitable access, and the potential misuse of advanced technologies.	Ethical guidelines for development and use, policies to ensure fair access and distribution, and public engagement to address societal concerns.

In the realm of medical applications, the idea of biocompatibility – defined as a substance's ability to coexist peacefully with living tissues without causing adverse reactions – is crucial. Due to their inventive qualities, smart materials typically require thorough assessments to determine their applicability in various biological environments [30]. In order to evaluate several factors, such as cytotoxicity, immunogenicity, and inflammatory response, extensive testing must be conducted. Even though many smart materials are very biocompatible, there may be difficulties due to intricate interactions with biological molecules, variations in material properties over time, and the possibility of byproduct release [31]. An important component that demands a full understanding of the progression of intelligent materials within the bodily systems over extended periods of time is the analysis of long-term effects. Degradation kinetics, steadiness, and the potential for decomposition products to accumulate are crucial factors that affect the efficacy and safety of smart materials [32]. The difficulty lies in extending these materials' useful life, stability, and biocompatibility beyond what is advised; this necessitates a combination of thorough testing, careful material design, and ongoing monitoring. Furthermore, due to the potential for prolonged exposure, it is crucial to conduct comprehensive assessments of the potential impacts of smart materials on tissues, organs, and overall systemic health as they become increasingly used in implanted devices and long-term therapies [33]. The active character of material-tissue interfaces should be taken into consideration when evaluating biocompatibility, rather than focusing solely on the first interactions between materials and tissues. Immune responses, tissue regeneration, and the potential for chronic inflammatory responses should all be included in this thorough evaluation. The thorough investigation of biocompatibility and long-term effects requires multidisciplinary cooperation between experts in the domains of biology, clinical practice, materials science, and regulatory bodies [34]. Predictive modeling, sophisticated in vitro and in vivo studies, and rigorous preclinical research must all be incorporated into a comprehensive evaluation of the safety and efficacy of smart materials. In order to validate the materials' effectiveness in real-world settings and clarify any potential long-term challenges, it is also important to carry out extensive clinical trials. Current advancements in information processing, wireless communication, micro- and nanotechnologies, and non-invasive biomedical measurement allow for smart miniaturization, body-worn sensors, transmission, processing, and non-invasive biomedical measurement [35]. Although the "smart health wearable" research and industry community is creating specific systems and applications to meet specific user needs, they still face some common critical issues, including biomedical sensors, scenarios of use, data security and confidentiality, risk analysis, user interface, medical knowledge/decision support, dissemination, user acceptance and awareness, business models, and exploitation [36].

While the field of smart materials holds immense promise for transforming medical and healthcare applications, overcoming these challenges is critical for their successful implementation [37]. Addressing biocompatibility, ensuring long-term stability, achieving scalability, navigating regulatory pathways, reducing costs, and integrating with existing systems are key areas that require ongoing research, innovation, and collaboration. By tackling these challenges, the prospective of smart materials to intensify patient care, improve therapeutic outcomes, and drive medical innovation can be completely realized [38].

4 Future Opportunities

The combination of bioscience and nanotechnology is creating a powerful collaboration that is transforming the biomedical industry. The ability to regulate material properties and interactions to an unprecedented degree is made possible by nanotechnology, which functions at the microscopic level. The mixture of biomaterials that have been specially designed to interact with biological systems opens up a wealth of possibilities for the creation of ground-breaking medical innovations [39]. By carefully modifying materials at the nanoscale, nanotechnology allows for the control of their mechanical, chemical, and physical characteristics. Improved biocompatibility, stability, and bioactivity result from the integration of biomaterials with custom-made alteration, producing substances that intently mimic the native habitat of living body tissues. The application of nanoscale biomaterials, which are specifically made to carry and distribute medications with extraordinary accuracy, has greatly enhanced the science of drug administration [40]. Pharmaceutical chemicals can be encapsulated by nanoparticles, nanocarriers, and nanogels, which allow for regulated release systems that provide accurate dosage administration and targeted distribution. This ability maximizes the effectiveness of treatment while lowering the likelihood of systemic side effects.

The field of smart (self-healing, multidimensional, compatible, responsive, and scalable) materials is rapidly evolving, presenting numerous future opportunities to revolutionize medical and healthcare applications [41]. It is anticipated that these materials will provide more complex answers to medical problems as research advances, enhancing patient outcomes and the effectiveness of healthcare. The following significant prospects are clearly illustrated and described in Fig. 2 and Tab. 3:

With the use of nanotechnology, biomimetic compositions that faithfully imitate the complex properties of organic tissues may be created. It is possible to construct these structures in a way that makes it easier to promote cell attachment, expansion, and divergence [42]. The incorporation of signals at the nanoscale into biomaterials is essential for making them essential in the domains of tissue engineering and regenerative medicine by stimulating tissue regeneration and controlling cellular responses. Utilizing biomaterials enabled by nanotechnology offers novel possibilities for imaging and diagnostic applications [43]. When used as contrast agents, functionalized nanoparticles can improve the visibility of specific tissues or biomarkers during imaging procedures. As a result, this phenomenon leads to improved monitoring of therapy efficacy and early illness identification. The creation of biomaterials with exact control over microscale properties that enable them to react selectively to anatomical cues is the basis for targeted therapeutics [44]. This responsiveness's potential can be used to speed up the creation of certain medications that are only activated in response to altered physiological conditions or illness markers, hence lowering the likelihood of side effects [45]. Table 3 shows a detailed prospect of smart materials in the healthcare sector.

The seamless integration of smart materials with microfluidics makes it easier to construct cutting-edge medical equipment and diagnostic systems. Because of their inherent adaptability, smart materials are very useful for controlling and manipulating fluid flow in microscale research [46]. Incorporating responsive properties in smart materials enable immediate signal detection, thereby increasing the sensitivity and accuracy of microfluidic devices. The cooperation of many components drives the development of lab-on-a-chip systems and small-scale diagnostic instruments, both of which seek to provide everyone access to high-tech medical testing. Combining cutting-edge technology with smart materials offers a number of advantages and disadvantages [47]. It is critical to consider the following: resolving any unfavorable interactions, optimizing production processes, and guaranteeing compatibility between smart materials and these technologies. Furthermore, due to ethical and regulatory concerns – such as the safety of integrated devices and the possible long-term effects of this technology – gene editing must be thoroughly examined [48]. In the event of an accident, mobile apps will allow the network to geographically connect to the nearest hospital and send data before admittance, potentially saving lives. In the future, new self-care techniques enabled by digitally improved technology are expected to be

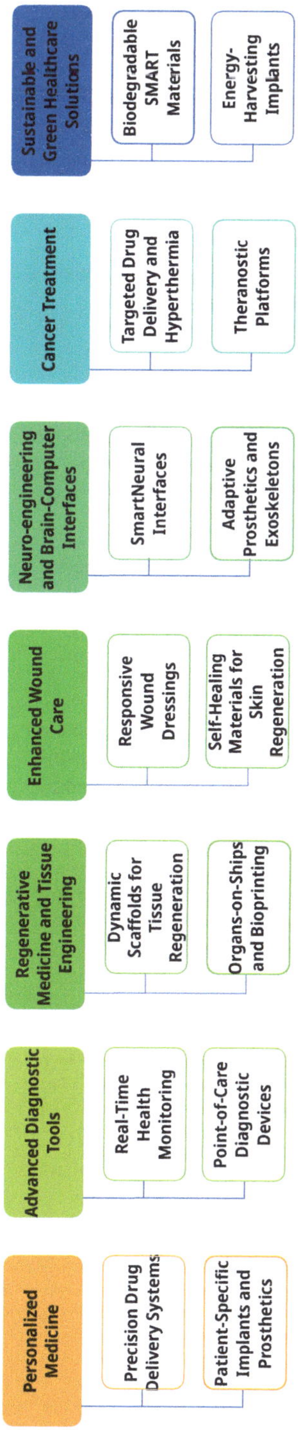

Fig. 2: Future opportunities for smart materials in healthcare sectors.

Tab. 3: Various sectors of healthcare depicting the specialization, opportunities, and future directions of smart materials.

Healthcare and medical sector	Specialization	Opportunity	Future directions
Personalized medicine	Precision drug delivery systems	Creating medication delivery systems that can be customized to meet each patient's unique needs in order to provide the best possible therapeutic results with the fewest possible negative effects.	SMART materials that respond to patient-specific biological markers, allowing for customized drug-release profiles.
	Patient-specific implants and prosthetics	Creating implants and prosthetics that adapt to the unique anatomical and physiological characteristics of each patient.	3D-printed SMART materials that can be customized in shape, size, and functionality for individual patients.
Advanced diagnostic tools	Real-time health monitoring	Development of wearable and implantable sensors that provide continuous, real-time monitoring of vital signs and biochemical markers.	SMART materials that integrate with digital health platforms, enabling remote monitoring and early detection of health issues.
	Point-of-care diagnostic devices	Creating portable diagnostic devices that use SMART materials to provide rapid and accurate diagnostics at the point of care.	Materials that change properties (e.g., color, fluorescence) in the presence of specific biomarkers, enabling quick and easy diagnosis in non-laboratory settings.
Regenerative medicine and tissue engineering	Dynamic scaffolds for tissue regeneration	Development of scaffolds that can dynamically interact with their environment to support and enhance tissue regeneration.	SMART scaffolds that release growth factors in response to cellular signals, promote vascularization, and degrade at controlled rates to match tissue growth.
	Organs-on-chips and bioprinting	Advancing organ-on-chip technologies and bioprinting using SMART materials to create functional tissue models for research and transplantation.	Using responsive materials to create more realistic and functional tissue constructs that can mimic the complex behaviors of human organs.

Tab. 3 (continued)

Healthcare and medical sector	**Specialization**	**Opportunity**	**Future directions**
Enhanced wound care	Responsive wound dressings	Development of wound dressings that respond to changes in the wound environment, such as infections or moisture levels.	Dressings that release antimicrobial agents, growth factors, or other therapeutic agents in response to specific wound conditions.
	Self-healing materials for skin regeneration	Creating materials that not only protect wounds but also actively participate in the healing process by mimicking natural skin regeneration.	Self-healing hydrogels and films that can repair themselves and promote faster, more effective wound healing.
Neuro-engineering and brain-computer interfaces	Smart neural interfaces	Developing interfaces that can dynamically interact with neural tissue to record and stimulate neural activity.	Materials that can adjust their properties in response to neural signals, providing more precise control and communication with neural tissues.
	Adaptive prosthetics and exoskeletons	Creating prosthetics and exoskeletons that can adapt to the user's movements and intentions, thereby improving functionality and comfort.	Using adaptive materials that can change stiffness and shape in real-time based on feedback from the user's nervous system.
Cancer treatment	Targeted drug delivery and hyperthermia	Enhancing the specificity and effectiveness of cancer treatments through targeted drug delivery and hyperthermia.	SMART nanoparticles that can home in on tumor cells, deliver chemotherapy drugs, and generate localized heat to kill cancer cells.
	Theranostic platforms	Combining therapeutic and diagnostic functions in a single platform to enable simultaneous treatment and monitoring of cancer.	Multifunctional materials that can diagnose the presence of cancer, deliver therapy, and monitor treatment responses in real-time.
Sustainable and green healthcare solutions	Biodegradable SMART materials	Developing environmentally friendly, biodegradable materials for medical applications to reduce medical waste and environmental impact.	Creating SMART materials that degrade safely in the body or environment, minimizing long-term waste and pollution.
	Energy-harvesting implants	Creating implants that can harvest energy from the body to power themselves, reducing the need for external power sources and surgeries to replace batteries.	Materials that convert mechanical, thermal, or biochemical energy from the body into electrical energy.

employed. For example, consider flash or continuous glucose monitoring systems, in which sensors can wirelessly send glucose data to a reader or mobile application, serve as display screens, and sound an alert in the event of hypo- or hyperglycemia [49]. Through the process of usability assessments, eye-tracking technology may be utilized to direct the patient-centered product development of innovative medication administration systems (self-injection devices). This advancement has the potential to significantly enhance the interaction between the connected system's parts – such as sensors, infusion pumps, remote controls, and mobile apps – improving the overall effectiveness of the product [50].

The future of smart materials in medical and healthcare applications holds immense promise for advancing personalized medicine, enhancing diagnostics, improving regenerative medicine, and developing more effective and sustainable treatments. Overcoming current challenges and continuing to innovate in this field will unlock new possibilities for improving patient care and healthcare outcomes [51]. Continued interdisciplinary research, collaboration, and investment in smart materials will be crucial in realizing these opportunities.

5 Conclusion

Investigating smart materials for biomedical applications reveals a landscape rich in possibilities, creative challenges, and opportunities. These carefully crafted materials, which take their cues from natural events, have the power to change the parameters of modern medical treatment. It is expected that they will usher in a new epoch of highly customized, highly responsive, and extraordinarily effective medical care.

The evolution of smart materials has demonstrated their prospective to significantly alter a number of biomedical fields. Numerous novel approaches to complex medical problems have been developed as a result of the wide variety of breakthroughs in materials science. Among these are materials with self-healing properties that mimic natural regeneration procedures and multipurpose materials that successfully include treatments and diagnostics. These technological advancements provide distinct chances to address challenging medical problems. The delivery of medications and the engineering of tissues may be highly controlled using materials that can adjust and react to environmental stimuli. Furthermore, wearable technology may be used to monitor health in real time with responsive materials. The changeable features of smart materials are a result of their better biocompatibility and integration, which expands their future applications in the realms of prostheses, implants, and personalized medicine.

However, there are several challenges in the way of this advancement. Establishing biocompatibility, understanding long-term effects, and improving self-healing mechanisms require rigorous and complete interdisciplinary collaboration. In order for monitoring features, therapies, and diagnostics to be successfully integrated, a high degree of engineering proficiency and close attention to legal requirements. Smart products must be successfully moved from laboratory settings to clinical applications, which necessitate cautiously navigating regulatory procedures and paying close attention to safety and ethical issues. Scalability, cost-effectiveness, and mass production factors highlight the necessity of accessible healthcare solutions. When smart materials are combined with emerging technologies like nanotechnology, 3D printing, gene editing, and microfluidics, their impact is increased as they keep getting better. These intersections have the power to transform healthcare practices and enable more personalized patient care. The development of smart materials has significantly advanced the field of medical and healthcare applications. These materials, designed to interact dynamically with their environment, offer innovative solutions that improve diagnosis, treatment, and patient care. Here, we discuss the latest advancements and their impacts in various medical and healthcare applications.

References

[1] Kumar A, Datta A, Kumar A, Kumar A. Recent advancements and future trends in next-generation materials for biomedical applications. In: Advanced Materials for Biomedical Applications. CRC Press; 2022 Dec 13. p. 1–19.

[2] Vashist A, Kaushik A, Ghosal Boca Raton A, Nikkhah-Moshaie R, Vashist A, Dev Jayant R, Nair M Journey of hydrogels to nanogels: a decade after.

[3] Kandhare AD, Thakurdesai PA, Wangikar P, Bodhankar SL. A Systematic Literature Review of Fenugreek Seed Toxicity by Using ToxRTool: Evidence from Preclinical and Clinical Studies. Heliyon. 2019 Apr 1; 5(4).

[4] Awasthi A, Saxena KK, Arun V. Sustainability and survivability in manufacturing sector. In: InModern Manufacturing Processes. Woodhead Publishing; 2020 Jan 1. p. 205–19.

[5] Ghosh A, Sharma A, Gupta N. Smart Materials Advancements, Applications and Challenges in the Shift to Industry 4.0. Ind 4.0 Clim Change. 2022 Oct 17; 177–88.

[6] Fernandes C, Heggannavar GB, Kariduraganavar MY, Mitchell GR, Alves N, Morouço P. Smart Materials for Biomedical Applications: The Usefulness of Shape-Memory Polymers. AMM. 2019 May 10; 890: 237–47.

[7] Arif ZU, Khalid MY, Zolfagharian A, Bodaghi M. 4D Bioprinting of Smart Polymers for Biomedical Applications: Recent Progress, Challenges, and Future Perspectives. React Funct Polym. 2022 Oct 1; 179: 105374.

[8] Castro NJ, Meinert C, Levett P, Hutmacher DW. Current Developments in Multifunctional Smart Materials for 3D/4D Bioprinting. Curr Opin Biomed Eng. 2017 Jun 1; 2: 67–75.

[9] Awasthi A, Gupta A, Saxena KK, Dwivedi RK, Kundalkar D, Abdul-Zahra DS, Joshi A, Saggu HS. Design and Analysis of Equal-Channel Angular Pressing of Al6061: a Comparative Study. Adv Mater Process Technol. 2024 Apr 2; 10(2): 408–17.

[10] Yilmaz ND. Multicomponent, semi-interpenetrating-polymer-network and interpenetrating-polymer-network hydrogels: smart materials for biomedical applications. In: Functional Biopolymers. Cham: Springer International Publishing; 2017 Oct 26. p. 281–342.

[11] Kumari CU, Murthy AS, Prasanna BL, Reddy MP, Panigrahy AK. An automated detection of heart arrhythmias using machine learning technique: SVM. In: Materials Today: Proceedings. 2021 Jan 1, vol. 45. p. 1393–98.
[12] Bahadur PS, Srivastava R, Srivastava V. Review on the Development of Smart Materials, Including Shape Memory Alloys, Characterization, and Recent Applications. Model Charact Process Smart Mater. 2023; 107–39.
[13] Srivastava R, Srivastava VK. Changing Scenario of Manufacturing Industries Due to Covid 19. Int Jl for Modern Trends in Sci Technol. 2020; 6(10): 143–46.
[14] Smith RC. Smart Material Systems: Model Development. SIAM. 2005 Jan 1.
[15] Leo DJ. Engineering Analysis of Smart Material Systems. John Wiley & Sons; 2007 Sep 10.
[16] Qader IN, Mediha KÖ, Dagdelen F, Aydoğdu Y. A Review of Smart Materials: Researches and Applications. El-Cezeri. 2019 Sep; 6(3): 755–88.
[17] Newnham RE, Ruschau GR. Smart Electroceramics. J Am Ceram Soc. 1991 Mar; 74(3): 463–80.
[18] Schwartz M. Smart Materials. CRC press; 2008 Nov 20.
[19] Aissa B, Memon NK, Ali A, Khraisheh MK. Recent Progress in the Growth and Applications of Graphene as a Smart Material: A Review. Front Mater. 2015 Sep 22; 2: 58.
[20] Akhras G. Smart Materials and Smart Systems for the Future. Can Mil J. 2000; 1(3): 25–31.
[21] Cao W, Cudney HH, Waser R. Smart Materials and Structures. PNAS. 1999 Jul 20; 96(15): 8330–31. Abdullah YS, Al-Alwan HA. Smart material systems and adaptiveness in architecture. Ain Shams Engineering Journal. 2019 Sep 1;10(3):623–38.
[22] Kim J, Yun S, Ounaies Z. Discovery of Cellulose as a Smart Material. Macromolecules. 2006 Jun 13; 39(12): 4202–06.
[23] Cao W, Cudney HH, Waser R. Smart Materials and Structures. PNAS. 1999 Jul 20; 96(15): 8330–31.
[24] Das L, Thakur A, Anand P. Application of MEMS sensors in IoT-integrated frameworks. In: Advances in MEMS and Microfluidic Systems. IGI Global; 2023. p. 153–72.
[25] Zieliński P. Smart cities as a sustainable development tool in spatial planning acts of the European Union and Japan: Comparative analysis. In: The Energy Transition in Japan. Routledge; 2025. p. 137–53.
[26] Pokines BJ, Garcia E. A Smart Material Microamplification Mechanism Fabricated Using LIGA. Smart Mater Struct. 1998 Feb 1; 7(1): 105.
[27] Qader IN, Kök M, Dagdelen F, Aydoğdu Y. A Review of Smart Materials: Researches and Applications. El-Cezeri. 2019 Sep; 6(3): 755–88.
[28] Bahl S, Nagar H, Singh I, Sehgal S. Smart materials types, properties and applications: A review. In: Materials Today: Proceedings. 2020 Jan 1, vol. 28. p. 1302–06.
[29] Bogue R. Smart Materials: A Review of Recent Developments. Assembly Autom. 2012 Feb 17; 32(1): 3–7.
[30] Thakur A, Vates UK, Mishra S. 4D printing using fused deposited shape memory polymer PLA: A state-of-art review. In: In2022 2nd International Conference on Technological Advancements in Computational Sciences (ICTACS). IEEE; 2022 Oct 10. p. 462–68.
[31] Agarwal T, Hann SY, Chiesa I, Cui H, Celikkin N, Micalizzi S, Barbetta A, Costantini M, Esworthy T, Zhang LG, De Maria C. 4D Printing in Biomedical Applications: Emerging Trends and Technologies. J Mater Chem B. 2021; 9(37): 7608–32.
[32] Rehman M, Yanen W, Mushtaq RT, Ishfaq K, Zahoor S, Ahmed A, Kumar MS, Gueyee T, Rahman MM, Sultana J. Additive Manufacturing for Biomedical Applications: A Review on Classification, Energy Consumption, and its Appreciable Role Since COVID-19 Pandemic. Prog Addit Manuf. 2023 Oct; 8(5): 1007–41.
[33] Naniz MA, Askari M, Zolfagharian A, Naniz MA, Bodaghi M. 4D Printing: A Cutting-Edge Platform for Biomedical Applications. Biomed Mater. 2022 Sep 26; 17(6): 062001.

[34] Korpi AG, Ţălu Ş, Bramowicz M, Arman A, Kulesza S, Pszczolkowski B, Jurečka S, Mardani M, Luna C, Balashabadi P, Rezaee S. Minkowski Functional Characterization and Fractal Analysis of Surfaces of Titanium Nitride Films. Mater Res Exp. 2019 Jun 12; 6(8): 086463
[35] Prateek S, Garg R, Kumar Saxena K, Srivastav VK, Vasudev H, Kumar N. Data-Driven Materials Science: Application of ML for Predicting Band Gap. Adv Mater Process Technol. 2023 Feb; 2: 1–0.
[36] Zhao W, Yue C, Liu L, Liu Y, Leng J. Research Progress of Shape Memory Polymer and 4D Printing in Biomedical Application. Adv Healthc Mater. 2023 Jun; 12(16): 2201975.
[37] Pon VD, Wilson KJ, Hariprasad K, Ganesh V, Ali HE, Algarni H, Yahia IS. Enhancement of Optoelectronic Properties of ZnO Thin Films by Al Doping for Photodetector Applications. Superlattices Microstruct. 2021 Mar 1; 151: 106790.
[38] Bhukya MN, Kota VR. A Quick and Effective MPPT Scheme for Solar Power Generation During Dynamic Weather and Partial Shaded Conditions. Eng Sci Technol Int J. 2019 Jun 1; 22(3): 869–84.
[39] Chauhan RB, Shah TV, Shah DH, Gohil TJ, Oza AD, Jajal B, Saxena KK. An Overview of Image Processing for Dental Diagnosis. Innov Emerg Technol. 2023 Jun 6; 10: 2330001.
[40] Shie MY, Shen YF, Astuti SD, Lee AK, Lin SH, Dwijaksara NL, Chen YW. Review of Polymeric Materials in 4D Printing Biomedical Applications. Polymers. 2019 Nov 12; 11(11): 1864.
[41] Joshi S, Rawat K, Karunakaran C, Rajamohan V, Mathew AT, Koziol K, Thakur VK, Balan AS. 4D Printing of Materials for the Future: Opportunities and Challenges. Appl Mater Today. 2020 Mar 1; 18: 100490.
[42] Navaz AN, Serhani MA, El Kassabi HT, Al-Qirim N, Ismail H. Trends, Technologies, and Key Challenges in Smart and Connected Healthcare. Ieee Access. 2021 May 11; 9: 74044–67.
[43] Touati F, Tabish R. U-Healthcare System: State-of-the-Art Review and Challenges. J Med Syst. 2013 Jun; 37: 1–20.
[44] Das L, Thakur A, Anand P. Application of MEMS sensors in IoT-integrated frameworks. In: Advances in MEMS and Microfluidic Systems. IGI Global; 2023. p. 153–72.
[45] Srivastava R, Bahadur PS, Jaiswal S, Kumar A, Srivastava AK. Combating high temperature challenges using pyrosic material. In: In2022 2nd International Conference on Innovative Practices in Technology and Management (ICIPTM). IEEE; 2022 Feb 23, vol. 2. p. 411–15.
[46] Bahadur PS, Jaiswal S, Srivastava R, Kumar A. Advanced application of nanotechnology in engineering. In: In2021 International Conference on Technological Advancements and Innovations (ICTAI). IEEE; 2021 Nov 10. p. 92–95.
[47] Thakur A, Vates UK, Mishra S. 4D printing using fused deposited shape memory polymer PLA: A state-of-art review. In: In2022 2nd International Conference on Technological Advancements in Computational Sciences (ICTACS). IEEE; 2022 Oct 10. p. 462–68.
[48] Rosso F, Marino G, Giordano A, Barbarisi M, Parmeggiani D, Barbarisi A. Smart Materials as Scaffolds for Tissue Engineering. J Cell Physiol. 2005 Jun; 203(3): 465–70.
[49] Khoo ZX, Teoh JE, Liu Y, Chua CK, Yang S, An J, Leong KF, Yeong WY. 3D Printing of Smart Materials: A Review on Recent Progress in 4D Printing. Virtual Phys Prototy. 2015 Jul 3; 10(3): 103–22.
[50] Srivastava R, Thakur A, Rana A. The Basic Understanding of Nano Materials: A Review. NM-BCI. 2024; 1–20.
[51] Zheng Y, Tang N, Omar R, Hu Z, Duong T, Wang J, Wu W, Haick H. Smart Materials Enabled with Artificial Intelligence for Healthcare Wearables. Adv Funct Mater. 2021 Dec; 31(51): 2105482.
[52] Jaglan A, Dogra N, Kumar P, Kumar A. Role of sustainable materials in COVID-19. In: Advances in Sustainable Materials. Elsevier; 2025 Jan 1. p. 455–69.
[53] Kapoor N, Bali A, Jatin J, Kumar P. Green Synthesis of Carbon Dots: A Path Toward Sustainability. Adv Sustain Mater. 2025 Jan; 1: 69–106.
[54] Patel S, John D, Kollarath RM, Kumar P. Sustainable materials for urban streets: trends, challenges, and case studies. In: Advances in Sustainable Materials. Elsevier; 2025 Jan 1. p. 515–30.

[55] Avani S, Sharma I, Nagpal S, Saini P, Kumar P, Kumar A. Applications of biopolymers in the agriculture and food industry. In: Advances in Sustainable Biomaterials. CRC Press; p. 323–54.
[56] Singh HV, Nagpal S, Mishra AK, Kumar A, Kumar P. The advent of biopolymers for potential applications in the medical and pharmaceuticals sector. In: Advances in Sustainable Biomaterials. CRC Press; p. 298–322.
[57] Tege S, Kumar P. Review of 4D Printing and Materials Enabling Industry 4.0 for Implementation in Manufacturing: An Indian context. 3D Print Technol: Digit Manuf, Artif Intell, Ind. 7(4): 143.
[58] Geeri S, Kolakoti A, Kumar P, Gudimetla A. Experimental and optimization analysis on mechanical properties of polymer-based natural composites. In: Advances in Sustainable Materials. Elsevier; 2025 Jan 1. p. 229–53.
[59] Yadav R, Praksah C, Kumar P, Kumar A. Identifying factors affecting the choice of sustainable materials: a factor analysis approach for sustainable development. In: Advances in Sustainable Materials. Elsevier; 2025 Jan 1. p. 289–304.
[60] Ogidi OI, Kumar P. Medical and pharmaceutical applications of sustainable biopolymers. In: Advances in Sustainable Materials. Elsevier; 2025 Jan 1. p. 431–53.
[61] Koushiki C, Christine J, Kumar P. Sustainable materials as a novel source for biomedical applications. In: Advances in Sustainable Materials. Elsevier; 2025 Jan 1. p. 471–85.
[62] Kumar A, Rajak DK, Kumar P, Kumar A, editors. Advances in Sustainable Biomaterials: Bioprocessing 4.0, Characterizations, and Applications. CRC Press; 2024 Oct 28.
[63] Yadav P, Kumar P, Saxena M, Dwivedi Y, Dubey A. 15 Application of Three-Dimensional Printing in Medical, Agriculture, Engineering, and Other Sectors. 3D Print Technol: Digit Manuf, Artif Intell, Ind 4 0. 2024 Jan 29; 311.
[64] Kumar A, Kumar P, Sebaey TA, Gambhir V, editors. Tools, Techniques, and Advancements in Engineering Materials Science. IGI Global; 2024 Dec 24.

Narinder Kumar*, Hardeep Singh, and Manjit Singh

Advancements in 3D Printing of Composite Materials: Challenges and Prospects

Abstract: Three-dimensional (3D) printing of composite materials has transformed production because it allows manufacturers to create robust weight-efficient complex structures which can be customized according to their needs. The present work explores 3D printing technology fundamentals and composite materials alongside their operational convergence with modern technology and material science developments. Continuous fiber-reinforced polymers remain a core development alongside hybrid manufacturing technologies and recycling methods that improve both sustainability and structural resilience. This technology faces broad implementation barriers due to material compatibility issues and economic challenges along with process reliability considerations. Current advancements in 3D printing capabilities have been driven by new discoveries in multifunctional composites and advanced computational tools and hybrid additive manufacturing processes. The future research outline requires deep exploration of self-healing materials together with sustainable resins and improved process control techniques to fully enable 3D-printed composites.

Keywords: 3D printing, composites, additive manufacturing, industrial applications

1 Introduction

Three-dimensional (3D) printing technology known as additive manufacturing (AM) has transformed complex and customized component creation across multiple industries [1]. 3D printing technology focuses on composite materials made from multiple phases because these materials possess enhanced mechanical thermal and electrical properties. The incorporation of composites with 3D printing systems allows manufacturers to develop purposeful lightweight components which demonstrate enhanced strength characteristics and application-specific versatility for aerospace, automotive, and healthcare and consumer needs. 3D printing materials made of composites enable

***Corresponding author: Narinder Kumar,** Department of Mechanical Engineering, Chandigarh Engineering College-CGC, Landran, Mohali, Punjab, India, e-mail: narinder.5282@cgc.edu.in
Hardeep Singh, Department of Mechanical Engineering, Chandigarh Engineering College-CGC, Landran, Mohali, Punjab, India, e-mail: hardeepsingh.5303@cgc.edu.in
Manjit Singh, Department of Mechanical Engineering, Chandigarh Engineering College-CGC, Landran, Mohali, Punjab, India, e-mail: manjit.3610@cgc.edu.in

https://doi.org/10.1515/9783111515717-012

enhanced control over properties when reinforcement elements are placed specifically to optimize orientation distribution within the structure. Engineers and designers benefit from this capability to produce parts with improved strength-to-weight ratios combined with better thermal resistance and increased durability relative to standard monolithic materials. The development of 3D printing technologies particularly fused filament fabrication (FFF) and stereolithography (SLA) and direct ink writing (DIW) has led to increased options for printable composite materials which now include polymer-matrix, ceramic-matrix, and metal-matrix composites [2]. Integration of 3D printing technologies with composite materials signals a transformative change in manufacturing, offering previously unimagined opportunities for innovation across industries. 3D printing involves the layer-by-layer fabrication of an object from digital models, enabling the possibility to fabricate complex designs while minimizing the waste in material. The capability is widely enhanced by the use of composite materials (materials consisting of two or more constituent materials) that have been engineered with the aim of achieving superior mechanical, thermal, or electrical properties. 3D printing works on the principle of layer-wise deposition of material using various techniques such as fused deposition modeling (FDM), SLA, selective laser sintering (SLS), and DIW technique differs from other techniques most strongly because of its material compatibility together with its precision levels and application ranges [3]. A digital model generated by CAD software becomes a 3D-coordinated blueprint which slicing algorithms transform into machine directives. 3D-printed composite materials incorporate thermoplastic polymers as their matrix material while adding fillers including carbon fibers, glass fibers, and nanoparticles. Matrix materials obtain superior properties through added reinforcements that yield improvements in tensile strength, stiffness, and thermal stability, for example, short-fiber composites function in filament-based printing and continuous-fiber reinforcements bring structural integrity demands. 3D printing within the framework of composite materials involves a thermoplastic polymer as the main material component supplemented by fillers consisting of carbon fibers alongside glass fibers and nanoparticles and other elements. Several characteristics improve within the matrix through integration of filler materials which boosts its tensile strength and rigidity and enhances thermal stability. Short-fiber composites dominate filament-based printing yet continuous-fiber reinforcement performs best for applications that need exceptional structural stability. Research and development continuously unfold new capabilities and applications space for this synergistic technology, providing solutions to some contemporary engineering challenges while being green [5, 6]. The industry stands on the verge of ground-breaking innovation, reshaping industries, and engendering new design methodologies. Besides, 3D printing with composites speeds up the process of prototyping and manufacture by reducing lead times and costs compared to traditional manufacturing techniques. It is this fast fabrication ability that may prove beneficial in industries that require iterative design and testing.

2 Fundamentals of 3D Printing and Composite Materials

2.1 Overview of 3D Printing Technologies

AM, the technologies of 3D printing, are processes where objects are fabricated layer by layer from digital models. Such technologies have achieved rapid production of highly customized and complex geometries, which are often impossible to produce with traditional manufacturing technologies. In this section, some key 3D printing technologies used in composite material manufacturing in terms of working principles, capabilities, and relevance to composite fabrication are explored.

- Fused filament fabrication (FFF): 3D printing using thermoplastic filaments through a heated nozzle which is used to create a part piece by layer is one of the most prevalent 3D printing technologies. Thermoplastic polymers are often reinforced with fibers, such as carbon or glass, to increase the mechanical properties for composites. Given its affordability and adaptation to make reinforced polymer matrix composites, FFF is particularly popular [7]. However, the issues are uniform reinforcement dispersion and optimal interfacial bonding.
- SLA: SLA cures liquid photopolymer resins layer by layer with a laser or digital light projector. SLA can be integrated with composite materials providing high resolution as well as more fine particles or fibers within the resin matrix, resulting in enhanced surface finish. This method is particularly appropriate for the fabrication of composites with complex geometry, however possibly with restriction to the use of compatible reinforcement with resin.
- SLS: Solid objects are created from powdered materials by a laser selectively sintering those materials, layer by layer, with SLS. The application of this technique to polymer and metal composites makes it possible for reinforcing particles or fibers to be directly incorporated into the powder bed. While high design flexibility and mechanical performance are offered across SLS, precise control of powder properties and process parameters is necessary to achieve part quality [8, 35].
- DIW: DIW is the extrusion of a viscous ink containing a mixture of polymers and reinforcements such as nanoparticles or short fibers through a nozzle. The ability to print custom composite materials with defined properties is provided by this technology. DIW as an ideal material to create functional composites with controlled microstructures for electronics and biomedical applications.
- Continuous Fiber 3D Printing (CF3D): Direct incorporation of continuous fibers into the printed structure is a major advance in composite manufacturing represented by CF3D. This method offers exceptional strength to weight ratio and directional control over the mechanical properties and is suitable for aerospace and automotive applications. CF3D is, however, a very sophisticated process, which

needs to be controlled to line up fibers at a specific angle and to produce pieces of constancy.

2.2 Composite Materials

Composite materials are the products of combining two or more different materials in a controlled fashion to produce a material better than the components [9]. Composites are typically a matrix that binds or supports reinforcing materials, whose use to provide strength, stiffness or other enhanced properties (Tab. 1). The result is a material stronger, lighter or more durable than its constituents that composites have become an integral part of modern manufacturing and engineering. Many structural composites and alloys were deformed by using incremental forming process [10, 11]. In this process a hemispherical tool has been used to deform the sheet which is made up of either an alloy or composite materials [12, 13].

Tab. 1: Types of composite materials.

Type	Explanation	Examples	Key applications
Polymer matrix composites (PMCs)	Composites with a polymer-based matrix reinforced with fibers like glass, carbon, or aramid.	Carbon fiber-reinforced polymer (CFRP), GFRP.	Aerospace and automotive
Metal matrix composites (MMCs)	Composites where a metal matrix is reinforced with ceramic or metallic particles or fibers.	Aluminum reinforced with silicon carbide.	Automotive, and thermal systems
Ceramic matrix composites (CMCs)	Composites with a ceramic matrix reinforced by ceramic or carbon fibers to improve toughness.	Silicon carbide fiber-reinforced silicon carbide	Aerospace and energy systems

Composite materials possess a range of desired properties that do suit their use in many ways, not the least of which being:

1. Higher strength-to-weight ratio: As with PMCs, composites typically use a lower weight to provide higher strength, which makes them ideal for use in applications where mobility and energy efficiency are demanded, to name but two: aerospace and automotive.
2. Thermal resistance: Critical for thermal shielding around energy systems, CMCs are known for their high-temperature tolerance preventing degradation.
3. Resistance to corrosion: While metals degrade in the environment, numerous composites are better able to resist the effects of the environment and are suitable for marine and chemical applications.

4. Adaptable properties: Composites are high freedom of design materials that can be effectively fabricated with a high freedom of design because of differing matrix and reinforcing materials to suit specific mechanical, thermal, or electrical requirements.
5. Durability: Long life in extreme applications can be engendered in composites by a high degree of corrosion, wear, fatigue, and environmental resistance.

Composite materials are versatile and used in so many industries. Lightweight composite materials save fuel while staying structurally sound in aerospace. In the automotive world, composites as a weight reduction tool and provide a way to improve vehicle durability. Biocompatible composite material finds medical application in implants and prosthetics. Concurrently the energy sector is using composites in wind turbines, battery components, and thermal shields.

3 Opportunities in 3D Printing of Composite Materials

AM – also called 3D printing – has made the manufacturing world upside down: Complex geometries can be created with unprecedented precision and efficiency. There is a significant frontier in 3D printing research in composite materials that combine two or more distinct phases to produce superior mechanical, thermal, or functional properties. However, it also enables further integration of these two technologies across many domains from aerospace to automotive to biomedical to construction industries respectively. 3D printing of composite materials is one of the primary opportunities due to material property tailoring for specific applications. Researchers engineer composites with exceptional strength to weight ratio, thermal stability, and corrosion resistance [14].

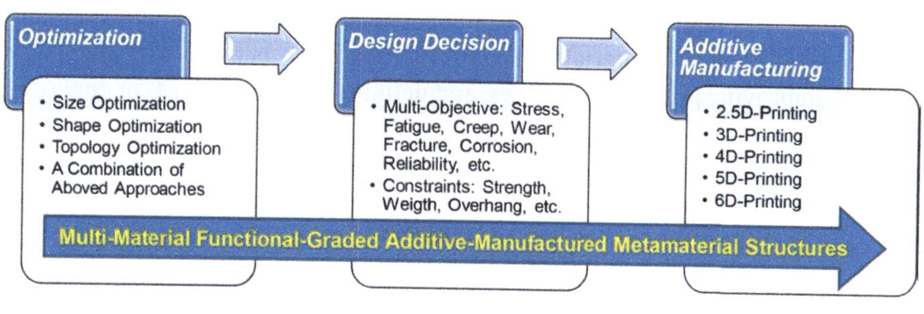

Fig. 1: Process algorithm of multi-material meta-material structures [15].

Combining materials such as polymers, ceramics, and metals with reinforcing fibers or nanoparticles [16]. It is particularly important to customize this 3D printing parameter in sectors such as aerospace since lightweight and durable components are a must since it allows the fabrication of intricate geometries that can rarely be obtained in traditional manufacturing techniques. When used on composite materials, this capability can be used to optimize the topologies that maximize material efficiency and performance. It has been observed that by printing lattice structures with reinforcing fibers provides high strength and stiffness at the same time as low material usage, resulting in both cost savings and improved sustainability [17].

Similarly, advances in multi-material printing technologies provide exciting opportunities for producing the functionally graded composites (Fig. 1). Spatially varying properties, for example, stiffness, conductivity, and thermal expansion, can be tailored in these materials to desired requirements [18]. In biomedical implants, where graded materials can mimic the transition between bone and soft tissue, this capability would be highly advantageous, as it will improve compatibility and functionality in such applications. The integration of digital design and simulation tools with 3D printing of composites enables researchers to predict and optimize material behavior before fabrication. By reducing the trial and error approach that is commonly used for composite manufacturing, computational models can simulate the impact of fiber orientation, layer thickness, and printing parameters and predict interplay between these factors with mechanical properties of the final product [8]. Traditional composite manufacturing often requires a lot of material removal and waste, while AM provides the ability to build components layer by layer using just the most needed material. Furthermore, composite formulations can be made with recycled materials or with bio-based reinforcements towards a circular economy [19]. Similar opportunities for 3D printing of composite materials exist in the biomedical field as the basis for patient specific solutions. Applications such as prosthetics, dental implants, and tissue engineering scaffolds can be easily tailored for biocompatible composites [20]. Complex geometries ensure better integration with biological systems, and these materials can provide long term mechanical strength and biocompatibility [21].

However, the current 3D printing of composite materials is constrained not only due to considerations of material formulation and process but also scalability. New materials, improving printing techniques, and defining robust standards of quality assurance should be addressed in future research. The advancements in these technologies will be significantly sped up by the real-time monitoring and optimization of the printing process by emerging technologies like artificial intelligence and machine learning [22].

4 Challenges in 3D Printing of Composite Materials

The potential advantages of 3D printing composite materials however are hampered by many difficulties that limit its widespread application and commercialization. In 3D printing, one of the major challenges is developing composite materials. Typically, poor dispersion of reinforcements, lack of material compatibility, and suboptimal interfacial bonding lead to reduced mechanical properties and inconsistent performance [23]. Moreover, feedstock materials suitable with specific printing process, for example, FDM and SLA, are limitedly available [19]. Several process related challenges arise in the complexities of printing composite materials (Fig. 2). For example, achieving desired mechanical properties necessitates precise control of fiber orientation in the print, but the state of the art in methods for controlling fiber orientation in the print is often poor [24]. Similarly, printed components suffer such issues as warping, delamination, and residual stress that could impact the quality and reliability of the printed component. Present 3D printing equipment has been designed for monolithic materials and may need a major modification to be able to handle composite materials. For instance, printers require nozzles (or extrusion system) tailored specifically to ink that can carry fiber reinforced polymers or high viscosity slurries [25]. However, barriers to the production of composite components through 3D printing exist, as the printing speeds are slow and the process of 3D printing is costly, leading to scalability issues for small scale/component and academic users, in particular. Although they make suitable prototypes and for small-scale manufacturing, these limitations prevent them from competitive with traditional techniques to produce in large volumes [16]. A large barrier to adoption of 3D-printed composites is the lack of standardized testing protocols and quality assurance. Despite sustainability benefits, the use of certain reinforcements like carbon nanotubes or nanoparticles raises environmental concerns and occupational health risks [26]; however, any change in material properties, process parameters, and environmental conditions may result in variability in the performance leading to inconsistent behavior and loss of confidence [27].

5 Recent Advances and Innovations

However, only in recent years have the scale of composite material manufacturing increased with the advent of 3D printing technology, enabling the printing of complex geometries and the tailoring of material properties. All of these contributions are responsible for what is happening today; more efficient and adaptable versions of production techniques. Continuous fiber-reinforced polymers (CFRPs) have emerged to provide increased mechanical properties, allowing the manufacture of relatively light and stiff structural constituents [28]. As a sustainable solution, recycling plastic waste with carbon fiber diminishes environmental consequence and preserves structural in-

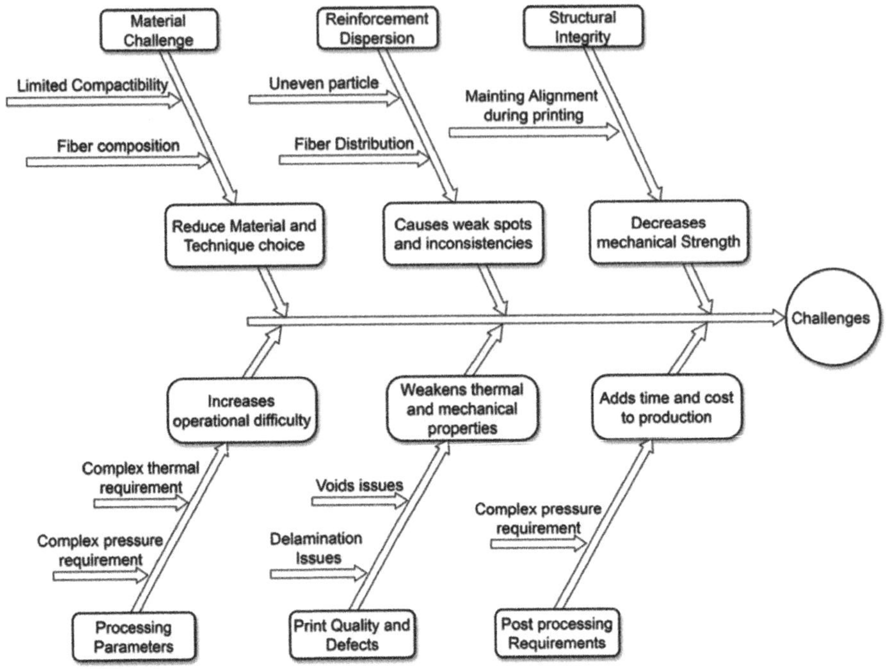

Fig. 2: Challenges in 3D printing of composite materials.

tegrity [29]. Achieving high precision and rapid processing speeds for thermoset polymer composites has been crucial in providing scalable manufacturing. Large, complex structures can be made, without the need for extensive post cure, using in-situ cure techniques used in conjunction with printing processes that have high production efficacy. Optimized 3D printing of composites narrows favored fiber distribution and topology, leading to increased predictability and performance in 3D printed composites. Although these advances present much scope for scalability, similar problems continue to exist with regard to consistency of mechanical performance and of materials and processes at our capability level.

An increased manufacturing capability through technological advancements in multi-material and hybrid 3D printing has enabled the printing of complex structures (Fig. 3). To produce stronger, more functional parts, these innovations allow integration of different materials and processes. Next, key improvements in this field are discussed: hybrid AM systems combining multiple AM processes and conventional manufacturing methods for improved efficiency and added capability. These advancements enable improved material properties, design flexibility, and the ability to make complex part in a single machine. AM that employs multiple techniques, that is, those of material removal and additive processes can be termed hybrid AM, thereby offering efficiency and part quality.

According to, hybrid AM is the first approach to successfully combine thermoplastics with photonic curable resins for stronger, printed parts that can be incorporated with reinforcing materials, thus further expanding multi-material and hybrid 3D printing. The range of applications is expanded with new methods for deposition of materials including thermoplastics and photo-curable resins. With technological improvements in multi-material and hybrid 3D printing scaling, this reduced constraints and enhanced geometric freedom. Based on this integration of aerodynamically focused nanoparticle printing micro-machining and focused ion beam milling, nanoscale manufacturing is enabled with broader material selection. Thus, full functional structures down to 50 nm in size are created through development of hybrid processes that integrate micro-machining and nanoparticle printing, capable of printing metals and ceramics. Technology improvements in multi-material 3D printing include the capacity to deposit more materials like a broader range of plastics and metals, or to integrate embedded electronics, allowing for complex structures with optical as well as electrical functionality [30]. New multi-material printing techniques permit the embedding of electronic circuits and sensors in printed objects for functionalities that other manufacturing does not possess. The early use of multiple materials in AM, highlighting the laminated object manufacturing process, which combined paper and resin to create composite objects, show casing foundational advancements in multi-material techniques within AM technologies. Prototyping and production: The increase in use of multi-material techniques to prototype is resulting in unique mechanical structures that improve product design and functionality.

6 Future Directions

Advancements in material science and manufacturing technologies result in abundant research opportunities in multifunctional composites and improved process control. They are areas in which significant potential for innovation lies in terms of adding capabilities and sustainability to composite materials such as in smart textiles (Fig. 4).

Driving advancements in technology and improving quality of life, the key research opportunities explored by include enhancing durability, scalability, and energy efficiency of multifunctional composites, optimizing fabrication techniques, material compatibility, and modeling multifunctional systems to realize reliable performance under various conditions [33]. Finally, research can integrate functionalities of self-healing, environmental sensing, or energy harvesting into textiles for their use in aerospace and healthcare. A review of additional work necessary to further improve functionalization techniques, improve 3D printing processes, and develop new material composites for multifunctional composites potential for innovation, particularly in enhancing the capabilities and sustainability of composite materials,[32]. Diels-

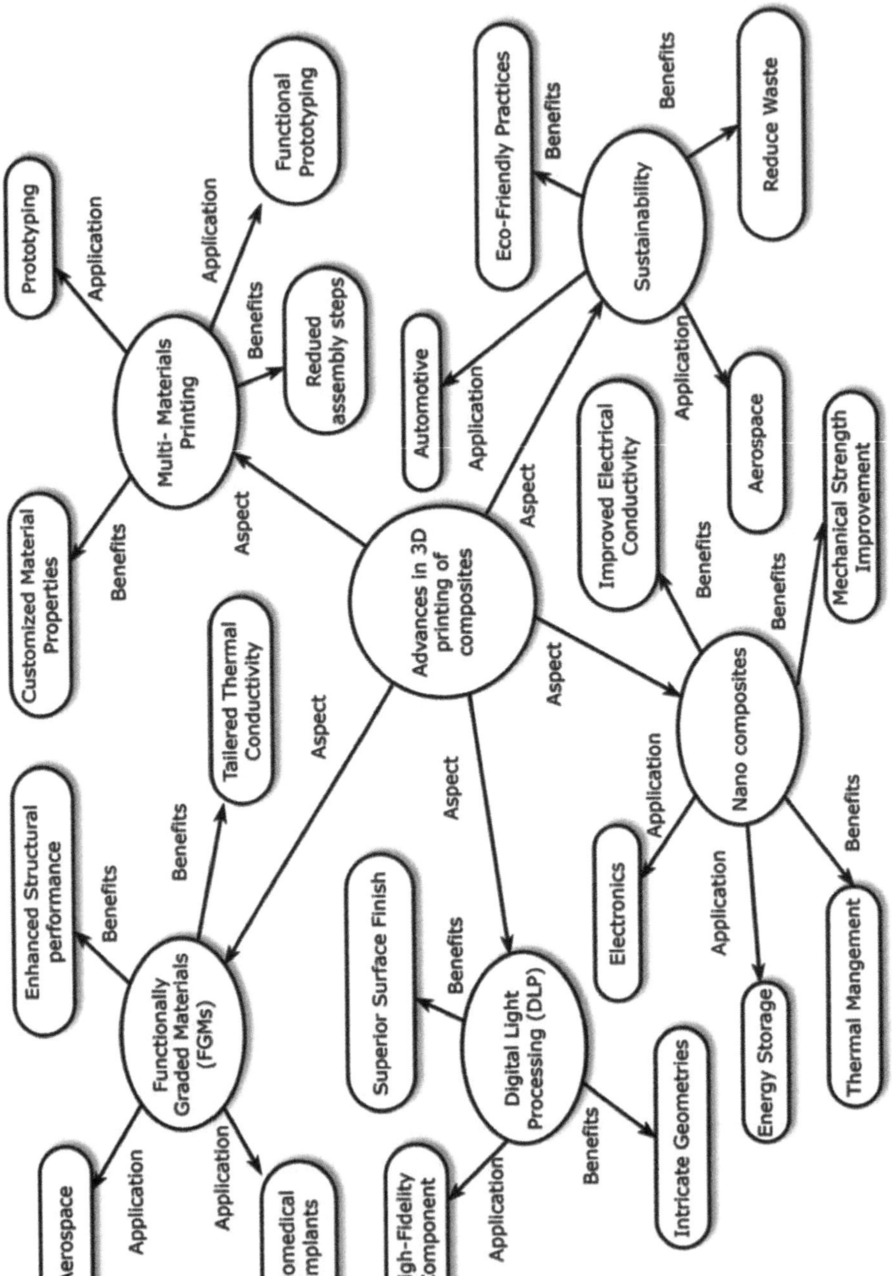

Fig. 3: Recent research highlights advances in 3D printing of composites.

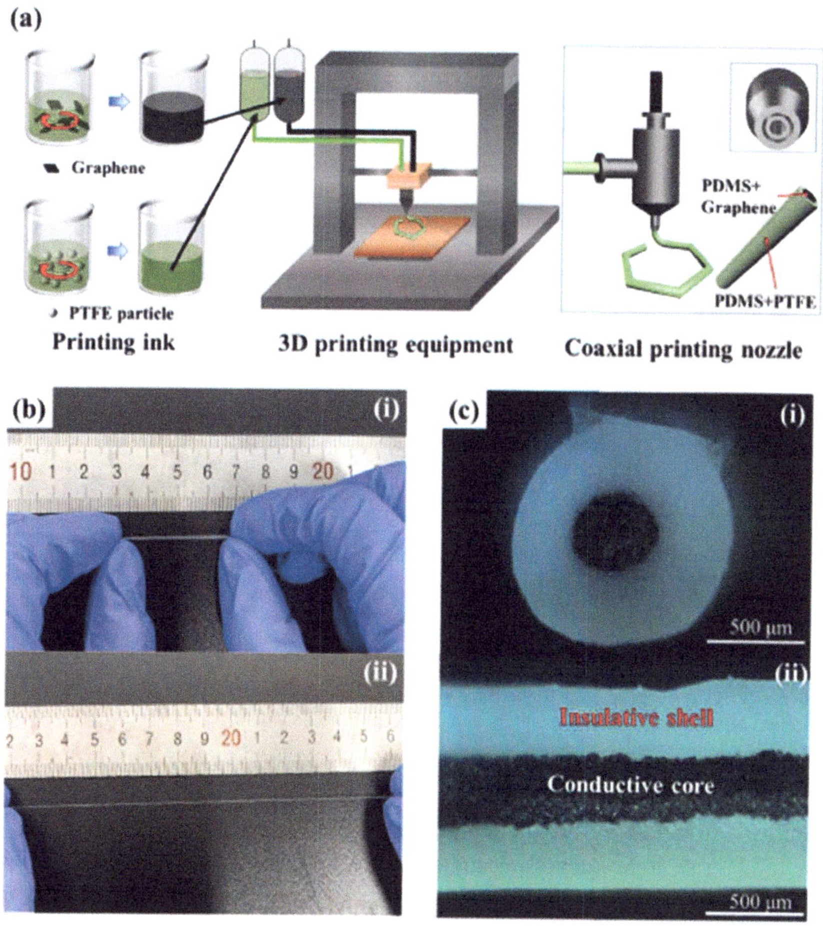

Fig. 4: Smart fibers that are coaxially stretchy are created using 3D printing technology. The coaxial stretchable smart fiber 3D printing technique is shown schematically in (a). (b) A physical representation of coaxial elastic fiber and an illustration of its extensibility; (c) coaxial elastic fiber optical microscopic pictures: (1) cross section and (2) longitudinal section [31].

Alder reactions are explored as a means toward the development of self-healing and reprocessable composites with the potential to produce innovative multifunctional materials (Fig. 5). Research opportunities to develop natural fiber composites, epoxy tooling via Industry 4.0, self-healing materials, mobile autonomous tools for process control, agility, and predictive simulation in composite manufacturing is still to be explored. Improving sustainability in composites manufacturing rests in research of natural fiber composites and ecofriendly resins. The development of mobile tools and predictive simulations can enhance process control, ensuring quality and efficiency in production. While these opportunities are promising, challenges such as material

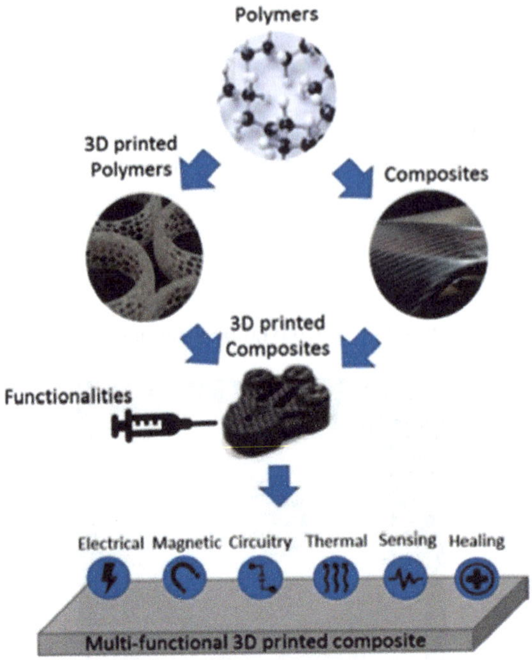

Fig. 5: Graphical representation of the multifunctional concept from polymers to 3D-printed composites [32].

compatibility and reliable performance under varying conditions remain critical areas for future research. Addressing these challenges will be essential for realizing the full potential of multifunctional composites and advanced manufacturing processes. These challenges necessitate multidisciplinary solutions, integrated advances in the field of materials science, process engineering, and computer modeling. The opportunities emerging technologies, such as artificial intelligence, real time monitoring systems, and advanced simulation tools, to break through these hurdles are to be studied in future study. Academia, industry, and regulatory bodies will also be collaborating to help establish standards and fast track the use of 3D-printed composites.

7 Conclusion

The integration of 3D printing into composite material manufacturing has ushered in new frontiers of manufacturing, which provide dramatic scalability, sustainability, and design flexibility. The state of the art in materials science and printing promises high-performance components for industries including aerospace, health care, and automotive. However, there is need to address problems of process consistency, material diversity, and economic efficiency for wide implementation. Future materials should be multifunctional, eco-friendly, and autonomous manufacturing tools. Further improvements in the utility and adaptability of 3D-printed composites are made possible by innovations in self-healing mechanisms, energy harvesting capabilities, and predictive simulations. Through resolving existing barriers, this approach could transform traditional manufacturing paradigms as well as create cost-effective and sustainable solutions in applications across the board.

References

[1] Alarifi IM. Revolutionising Fabrication Advances and Applications of 3D Printing with Composite Materials: A Review. Virtual Phys Prototyp. 2024; 19(1). doi: 10.1080/17452759.2024.2390504

[2] Alami AH, Ghani Olabi A, Alashkar A, Alasad S, Aljaghoub H, Rezk H, et al. Additive Manufacturing in the Aerospace and Automotive Industries: Recent Trends and Role in Achieving Sustainable Development Goals. Ain Shams Eng J. 2023; 14(11): 102516.

[3] Saroia J, Wang Y, Wei Q, Lei M, Li X, Guo Y, et al. A Review on 3D Printed Matrix Polymer Composites: Its Potential and Future Challenges. Int J Adv Manuf Technol. 2020; 106(5–6): 1695–721.

[4] Pervaiz S, Qureshi TA, Kashwani G, Kannan S. 3D Printing of Fiber-reinforced Plastic Composites Using Fused Deposition Modeling: A Status Review. Materials (Basel). 2021; 14(16). 4520. https://doi.org/10.3390/ma14164520

[5] Singh M, Singh C, Gangacharyulu D. Numerical Analysis of Unsteady Flow through Unsaturated Soil of Uniform Sized Spheres with Distinct Packing Arrangements. Mater Today Proc. 2022; 56: 2973–79. Available from https://www.sciencedirect.com/science/article/pii/S2214785321070887

[6] Singh M, Singh C, Gangacharyulu D. Numerical analysis on variations of thermal and hydrological properties during water flow through unsaturated soil BT. In: Saha P, Subbarao PMV, Sikarwar BS, editors. Advances in Fluid and Thermal Engineering. Singapore: Springer Singapore; 2019. p. 357–68.

[7] Ning F, Cong W, Qiu J, Wei J, Wang S. Additive Manufacturing of Carbon Fiber Reinforced Thermoplastic Composites Using Fused Deposition Modeling. Compos Part B Eng. 2015; 80: 369–78. Available from https://www.sciencedirect.com/science/article/pii/S1359836815003777

[8] Soodesh CY, Roy B. Natural fiber and nanoparticles reinforced natural fiber for structural composite applications. In: Singh SB, Gopalarathnam M, Kodur VKR, Matsagar VA, (eds.). Fiber Reinforced Polymeric Materials and Sustainable Structures [Internet]. Singapore: Springer Nature Singapore; 2023. p. 139–58. Available from https://doi.org/10.1007/978-981-19-8979-7_11

[9] Singh H, Singh J, Kumar S. Effect of Processing Conditions and Electrode Materials on the Surface Roughness of EDM-processed Hybrid Metal Matrix Composites. Int J Light Mater Manuf. 2024; 7(3): 480–93. Available from https://doi.org/10.1016/j.ijlmm.2023.12.001

[10] Kumar N, Lingam R, Agrawal A. Enhancement of Incremental Forming Process Formability by Using Improved Clamping and Multi Stage Deformation Strategies. Int J Adv Manuf Technol. 2023; 129: 659–70. Available from https://doi.org/10.1007/s00170-023-12298-w

[11] Kumar N, Mahala M, Agrawal A. A Coupled Thermo-mechanical Model for Warm Single-point Incremental Forming Process. Arch Civ Mech Eng [Internet]. 2025; 25(1). Available from https://doi.org/10.1007/s43452-024-01107-0

[12] Kumar N, Bharti S, Krishnaswamy H, Agrawal A. Exploring Deformation Mechanics of Temperature Assisted Incremental Forming with Hybrid Heating. J Manuf Process [Internet]. 2023; 104(427): 472–84. Available from https://doi.org/10.1016/j.jmapro.2023.09.017

[13] Kumar N, Upadhyay RK, Sharma CS. Surface Quality Enhancement of Warm Incremental Forming Process Using Solid Lubricant with Varying Etching Period. 2023; 427. Proceedings of the Institution of Mechanical Engineers, Part B: Journal of Engineering Manufacture. 2023; 238(9): 1327–1337.
[14] Goh GD, Yap YL, Agarwala S, Yeong WY. Recent Progress in Additive Manufacturing of Fiber Reinforced Polymer Composite. Adv Mater Technol [Internet]. 2019 Jan 1; 4(1): 1800271. Available from https://doi.org/10.1002/admt.201800271
[15] Esfarjani SM, Dadashi A, Azadi M. Topology Optimization of Additive-manufactured Metamaterial Structures: A Review Focused on Multi-material Types. Forces Mech. 2022 March; 7: 100100. Available from https://doi.org/10.1016/j.finmec.2022.100100
[16] Zhang J, An Q. Topology Optimization of Fibre Reinforced Polymer Lattice Structures for Additive Manufacturing. Compos Sci Technol. 2023; 242: 110144. Available from https://www.sciencedirect.com/science/article/pii/S0266353823002373
[17] Wen J, Zhou L, Ye T. Polymer Ionogels and Their Application in Flexible Ionic Devices. SmartMat. 2024; 5(2). doi:10.1002/smm2.1253
[18] Kumar MS, Farooq MU, Yang CH, Kavimani V, Adediran AA, Ross NS. Achieving Effective Interlayer Bonding of PLA Parts during the Material Extrusion Process with Enhanced Mechanical Properties. Sci Rep. 2023; 1–21. Available from https://doi.org/10.1038/s41598-023-33510-7
[19] Al Rashid A, Koç M. Additive Manufacturing for Sustainability and Circular Economy: Needs, Challenges, and Opportunities for 3D Printing of Recycled Polymeric Waste. Mater Today Sustain. 2023; 24: 100529. Available from https://doi.org/10.1016/j.mtsust.2023.100529
[20] Maintz M, Tourbier C, de Wild M, Cattin PC, Beyer M, Seiler D, et al. Patient-specific Implants Made of 3D Printed Bioresorbable Polymers at the Point-of-care: Material, Technology, and Scope of Surgical Application. 3D Print Med. 2024; 10(1): 1–10.
[21] Birniwa AH, Abdullahi SS, Ali M, Mohammad REA, Jagaba AH, Amran M, et al. Recent Trends in Treatment and Fabrication of Plant-Based Fiber-Reinforced Epoxy Composite: A Review. J Compos Sci. 2023; 7(3): 120. https://doi.org/10.3390/jcs7030120
[22] Hassan M, Misra M, Taylor GW, Mohanty AK. A Review of AI for Optimization of 3D Printing of Sustainable Polymers and Composites. Compos Part C Open Access. 2024 September 2024; 15.100513; https://doi.org/10.1016/j.jcomc.2024.100513
[23] Alharbi M, Kong I, Patel VI. Simulation of Uniaxial Stress–strain Response of 3D-printed Polylactic Acid by Nonlinear Finite Element Analysis. Appl Adhes Sci [Internet]. 2020; 8(1): 1–10. Available from https://doi.org/10.1186/s40563-020-00128-1
[24] Liakos IL, Mondini A, Del Dottore E, Filippeschi C, Pignatelli F, Mazzolai B. 3D Printed Composites from Heat Extruded Polycaprolactone/sodium Alginate Filaments and Their Heavy Metal Adsorption Properties. Mater Chem Front. 2020; 4(8): 2472–83.
[25] Arockiam AJ, Subramanian K, Padmanabhan RG, Selvaraj R, Bagal DK, Rajesh S. A Review on PLA with Different Fillers Used as A Filament in 3D Printing. Mater Today Proc. 2021 October; 50: 2057–64.
[26] Mamo HB, Adamiak M, Kunwar A. 3D Printed Biomedical Devices and Their Applications: A Review on State-of-the-art Technologies, Existing Challenges, and Future Perspectives. J Mech Behav Biomed Mater. 2023 May; 143: 105930. Available from https://doi.org/10.1016/j.jmbbm.2023.105930
[27] Sharafi Zamir S, Fathi B, Ajji A, Robert M, Elkoun S. Crystallinity and Gas Permeability of Poly (Lactic Acid)/Starch Nanocrystal Nanocomposite. Polymers (Basel). 2022; 14(14): 1–14.
[28] Yu K, Dunn ML. Design and Extrusion-Based 3D Printing of Continuous Fiber Composites: Status, Challenges, and Opportunities. Langmuir. 2024 Apr 30; 40(17): 8751–59. Available from https://doi.org/10.1021/acs.langmuir.4c00139
[29] Patel KS, Shah DB, Joshi SJ, Patel KM. Developments in 3D Printing of Carbon Fiber Reinforced Polymer Containing Recycled Plastic Waste: A Review. Clean Mater. 2023; 9(May): 100207. Available from https://doi.org/10.1016/j.clema.2023.100207

[30] Ready S, Whiting G, Ng TN, Alto P, Alto P. Multi-material 3D printing. 2014. p. 120–23. In Proc. IS&T Int'l Conf. on Digital Printing Technologies and Digital Fabrication (NIP30), 2014, pp120–123, https://doi.org/10.2352/ISSN.2169-4451.2014.30.1.art00031_1

[31] Chen Y, Deng Z, Ouyang R, Zheng R, Jiang Z, Bai H. Nano Energy 3D Printed Stretchable Smart Fibers and Textiles for Self-powered E-skin. Nano Energy. 2021 February; 84: 105866. https://doi.org/10.1016/j.nanoen.2021.105866

[32] Bekas DG, Hou Y, Liu Y, Panesar A. 3D Printing to Enable Multifunctionality in Polymer-based Composites: A Review. Compos Part B Eng. 2019; 179: 107540. Available from https://doi.org/10.1016/j.compositesb.2019.107540

[33] Chaudhary B, Winnard T, Oladipo B, Das S, Matos H. Review of Fiber-Reinforced Composite Structures with Multifunctional Capabilities through Smart Textiles. Textiles. 2024; 4(3): 391–416.

[34] Tege S, Kumar P. Review of 4D Printing and Materials Enabling Industry 4.0 For Implementation in Manufacturing: An Indian Context. 3D Print Technol: Digit Manuf, Artif Intell, Ind. 7(4): 143.

[35] Anand S, Satyarthi MK, Bharti PS, Kumar P, Kumar A. Digital Twin Integration for Enhanced Control in FDM 3D Printing. In Industry 4.0 Driven Manufacturing Technologies. Cham: Springer Nature Switzerland; 2024 Sep 14. p. 373–88.

Amit Kumar, Siddharth Yadav, Anuj Kumar Sharma, Santosh Kumar, and Govind Kumar Verma*

Colloidal Formulation and Robocasting of Silicon Nitride Scaffolds with Tailored TPMS Structures

Abstract: Silicon nitride (Si_3N_4), combined with alumina and yttria as liquid-phase sintering aids, is a promising bioceramic for orthopaedic and dental applications due to its biocompatibility, wear resistance, and mechanical reliability. Despite its advantages, the fabrication of complex, porous Si_3N_4 architectures with tailored mechanical and biological functionality remains challenging due to limitations in slurry stability, rheology control, and geometric accuracy during additive manufacturing. This study addresses this gap by developing a non-Newtonian, shear-thinning colloidal suspension containing ~42 vol.% solid loading of Si_3N_4 with sintering aids, explicitly optimized for robocasting. The ink formulation utilizes polyvinyl alcohol (PVA) as a binder, carboxymethyl cellulose (CMC) as a viscosifier, and glycerol as a gelation agent to achieve suitable rheological properties and shape fidelity. Different triply periodic minimal surface (TPMS) scaffolds were successfully fabricated using this optimized ink. Process parameters were fine-tuned for each geometry, resulting in printed parts with relative densities ranging from 56% to 92%. After debinding and liquid-phase sintering, the structures exhibited a fibrous β-Si_3N_4 microstructure and a fracture toughness of approximately ~5 $MPa \cdot m^{1/2}$. These findings demonstrate the potential of 3D-printed Si_3N_4 scaffolds with bioinspired TPMS architectures for patient-specific, load-bearing biomedical implants.

Keywords: Additive manufacturing, silicon nitride, robocasting, rheology, TPMS

***Corresponding author: Govind Kumar Verma,** Department of Mechanical Engineering, Indian Institute of Technology (BHU), Varanasi 221005, Uttar Pradesh, India, e-mail: govindkumarverma.rs.mec20@iitbhu.ac.in

Amit Kumar, Centre for Advanced Studies, Dr. A.P.J. Abdul Kalam Technical University, Lucknow 226031, Uttar Pradesh, India

Siddharth Yadav, Centre for Advanced Studies, Dr. A.P.J. Abdul Kalam Technical University, Lucknow 226031, Uttar Pradesh, India

Anuj Kumar Sharma, Centre for Advanced Studies, Dr. A.P.J. Abdul Kalam Technical University, Lucknow, 226031 Uttar Pradesh, India

Santosh Kumar Department of Mechanical Engineering, Indian Institute of Technology (BHU), Varanasi 221005, Uttar Pradesh, India

https://doi.org/10.1515/9783111515717-013

1 Introduction

Since the early 1970s, silicon nitride (Si_3N_4) has been extensively used in structural applications at ambient and elevated temperatures. This is primarily because of its remarkable qualities, which include hardness, superior wear resistance, stability at high temperatures, excellent mechanical strength, and resistance to thermal shock [1, 2]. Si_3N_4 is widely used in orthopaedic implants, dental prostheses, cardiovascular devices, ophthalmology, and reconstructive surgery because of its flexibility as a bioceramic [3, 4]. A controlled heat treatment can alter the microstructure of Si_3N_4 materials, resulting in elongated (flexible) grains that exhibit increased durability and toughness. Among the highest fracture toughness values recordable for ceramics are those associated with these in situ-toughened Si_3N_4 ceramics, with flexural strength exceeding 1 GPa [1, 5]. Silicon nitride's exceptionally flexural and compressive strength, remarkable fracture toughness, low-friction properties, and superior resistance to corrosion and wear all support these applications. Its excellent antibacterial efficacy against germs and viruses, high biocompatibility, and capacity to form a link with bone tissue all contribute to its increased appropriateness [2, 6, 31, 35]. Due to its advantageous properties, such as remarkable strength and resistance to fracture, spinal fusion implants have been clinically made using Si_3N_4, and various biomedical applications, including parts for dental prostheses and hip and knee joint replacements, are currently under development [7, 8].

In addition to being deemed significant for high-temperature mechanical applications, the near-net form manufacturing of Si_3N_4 components with customised geometry and internal architecture is considered highly advantageous to customised orthopaedic applications. The high hardness of Si_3N_4 frequently makes the post-processing of densified structures to obtain complicated geometries expensive. Complex forms are commonly needed for implants in biomedical settings to address bone abnormalities or other anatomical characteristics. The construction may be dense, porous, or have graduated porosity depending on the particular medical need. For instance, implants used in spinal restoration are frequently made with a porous core to promote bone ingrowth and biological integration and a thick outer shell to give the strength required. The form of the near net using conventional shaping methods including injection moulding [9], slip casting [10, 11], and gel casting [12, 13]. Si_3N_4 ceramics in both basic and intricate shapes have been created. In contrast, these techniques are less capable of controlling the fine distribution of porosity and pore size, which are necessary for successful implant integration with host bone. At the same time, additive manufacturing (AM), also known as three-dimensional (3D) printing, has demonstrated remarkable control over the interior microstructure and outward geometry of ceramic green bodies [32–34]. To reduce the manufacturing costs of Si_3N_4 implants and hasten their acceptance in biomedical applications, a pragmatic strategy is required to address the challenges associated with moulding and processing irregularly shaped and patient-specific Si_3N_4 bio ceramic components [14].

Direct ink writing (DIW), often called robocasting, is a material extrusion method. The development and optimization of oxide ceramic ink have been the focus of current research [15, 16]. Despite the lack of progress in developing non-oxide ceramic ink, polyethyleneimine was used as a binder and dispersion in formulating an aqueous Si_3N_4 paste for robocasting. This technique created porous and dense Si_3N_4 ceramics with four-point bending strengths of 89 ± 11 MPa and 552 ± 68 MPa [1]. During the printing process, the deposition route was optimized to remove inter-layer gaps in an Si_3N_4 paste with a solid concentration of up to 43 vol.%. As a result of this modification, the production of dense Si_3N_4 components (95% density) with bending strengths of more than 500 MPa was achieved [17]. Using different ratios of sintering additives, the biological activity of large-pore Si_3N_4 scaffolds (600–750 μm) created by DIW was evaluated. The biological response of the scaffold and the Si_3N_4 grain boundary phase composition was closely related. The 10 mol% Y_2O_3-Al_2O_3-SiO_2-doped Si_3N_4 (SN10YAS) composite exhibited the highest degree of biological activity and was identified by the notable presence of yttrium aluminium silicate ($Y_2Al_2SiO_7$) glass phase at the grain boundaries [6]. Regarding DIW, recent research highlighted a breakthrough where scientists used an aqueous colloidal ink containing 44% solids by volume (with around 12% sintering aids by weight) along with a small amount of gelling and dispersing agents. This approach showed great promise, successfully producing bulk Si_3N_4 samples with flexural strength and densities reaching nearly 97% of the theoretical maximum. However, the mechanical characteristics were not provided. Micro-computed tomography scans discovered periodic porosity at the filament interfaces despite the high density. This is a typical problem in aligned print patterns that results from insufficient filling [18]. To tackle the issue of particle instability in Si_3N_4 water-based suspensions, commercially available comb-type polymer concrete superplasticizers commonly used in the concrete industry were employed to stabilize these aqueous suspensions, even at high solids loading levels of up to 45 vol.%. It was investigated how the kind and concentration of polymer dispersants and the addition of polymer binders affected the mechanical and rheological characteristics. The created slurries and the findings of these experiments will serve as a reference for this investigation [19, 20].

Robocasting is an AM technique that uses a thick, concentrated slurry or paste, which is carefully pushed out layer by layer through a nozzle usually between 100 μm and 1 mm wide, following precise computer guidance to build up complex shapes [21]. For a variety of ceramics, including structural ceramics like silicon carbide and alumina [22]. Robocasting has been used to create green bodies and bioceramics like hydroxyapatite, beta-tricalcium phosphate [23–25], and bioactive glass [26–28], with a pH maintained between 7.8 and 8.5. Robocasting has also been applied to make Si_3N_4 parts using water-based slurries that are packed with a high amount of solid material, about 52% by volume of Si_3N_4 particles averaging 0.77 μm in size. The mixture includes a small amount of a flocculating agent (0.4% aluminium nitrate) and a low concentration (1%) of an anionic polyelectrolyte called Darvon 821A to help keep everything well-dispersed [29]. After sintering, the parts shrank by about 16% in size and reached a final density of 3.21 g/cm^3, which is nearly 99% of the theoretical maximum. Four-point bending tests on these sintered samples showed an impressive flexural strength of 737 ± 38 MPa, highlighting their strong and reliable performance [1].

This research aims to create a Si_3N_4 paste with a high solid content and optimise robocasting parameters to produce ceramic components free of flaws. The commercial Si_3N_4 nano-powder was chosen because of its good sintering behaviour, despite dispersion issues. Using Isobam104 as a dispersant, a stable, gel-forming paste that dries at room temperature without breaking was made possible. DIW was used to produce Si_3N_4 components with various topologies. Needle size, spacing, and angle measurements were used to regulate the macropore forms and diameters. In addition to hot isostatic pressing (HIP), the green portions underwent drying, debinding, and sintering. Interfilamentous bonding and defect development were examined using two nozzle diameters. The drying behaviour and rheological characteristics were investigated in a controlled humidity environment to minimise structural defects. Insights for the optimisation of DIW in advanced Si_3N_4 ceramic fabrication were gained by evaluating and linking the mechanical characteristics of the printed samples with their microstructure using scanning electron microscopy (SEM) and X-ray CT analysis.

2 Experimental Procedure

2.1 Feedstocks

In this research, commercially available Si_3N_4 powder with a molecular weight (M_w) of 140.28 g/mol was used; it had an average particle size of about 1 µm, a high purity of 99.6%, and a density of 3.44 g/mL. The powder also had an average particle size (D_{50}) of 1.70 µm and was primarily in the β-phase. Additionally, aluminium oxide (Al_2O_3) nanoparticles (M_w = 101.96 g/mol) with particle sizes under 100 nm, purity of 99.9%, density of 3.98 g/cm^3, as well as yttrium oxide (Y_2O_3) nanoparticles (M_w = 225.8 g/mol) with particles smaller than 50 nm and density of 5.01 g/cm^3, were sourced from Nano Research Elements in the UK. For the polymer components, PVA with an average $M_W \approx$ 8,000–10,000, density between ~1.19 and 1.31 g/cm^3, and polyethylene glycol (average $M_w \approx$ 400 g/mol), and hydroxypropyl methylcellulose (HPMC) powder (M_w 80,000 g/mol, particle size 20 mesh) were purchased from Sigma Aldrich, Canada. Distilled water (H_2O), with a $M_w \approx$ 18.02 g/mol, was sourced locally from Delhi, India, and carboxymethyl cellulose (CMC) was obtained from a supplier in Mumbai, India.

2.2 Preparation of Paste

The paste for the $Si_3N_4/Al_2O_3/Y_2O_3$/PVA composite was prepared through a carefully controlled process to ensure uniformity, stability, and optimal material performance, as shown in Fig. 1. Initially, composition ratios reported in the literature and supported by preliminary investigations, demonstrating improved fluid absorption and enhanced mechanical properties compared to pure Si_3N_4, were used to determine the required powder quantities. Accordingly, 9.0 g of Si_3N_4, 0.4 g of Al_2O_3, and 0.6 g of Y_2O_3 powders were accurately weighed. The powders were then thoroughly blended using a centrifuge mixer machine, by which homogeneity and uniform dispersion were achieved. This mixing process was considered essential, as interparticle interactions were enhanced, agglomeration was minimized, and sintering behaviour was improved. After mixing, the ceramic powder blend was reserved for subsequent incorporation into the binder solution.

In the next step, 0.2 g of PVA was precisely weighed and dissolved in 20 mL of HPLC-grade solvent. The dissolution was carried out under magnetic stirring at 80 °C for 1 h to ensure complete solubilization. Following this, 0.1 g of CMC was added to the solution and allowed to dissolve at room temperature over a period of 18 h with continuous magnetic stirring. Through this prolonged mixing, a homogeneous polymer solution was obtained, which was considered essential for maintaining the stability and performance of the composite. The dissolution of PVA and CMC significantly improved the compatibility and interaction between the components, providing a stable matrix for the integration of ceramic particles. After the binder solution had been prepared, the pre-mixed ceramic powders (Si_3N_4, Al_2O_3, and Y_2O_3) were gradually added to the polymer solution in small increments. The resulting slurry was subsequently mixed using the centrifuge mixer machine to ensure uniform dispersion of ceramic particles within the polymer matrix. A homogeneous ceramic-polymer paste with consistent rheological and compositional properties was thereby obtained, making it suitable for further forming and robocasting processes. Table 1 summarizes the composition of the formulated pastes employed in the DIW process.

2.3 Robocasting

Green components were made using robocasting. The layer-by-layer construction of 3D structures was precisely managed through independent control of the *x*, *y*, and *z* axes, using computer-aided software (Hyrel3d, Hydra 16AS Standard, AMS-India). To minimize interference during printing, the ceramic paste was loaded into a 25-mL syringe and gently tapped to eliminate any air bubbles. The paste was then extruded onto a vaseline-coated glass slide using a three-axis gantry system controlled by a multi-axis motion controller on the Hyrel3d platform. Vaseline was used as a release

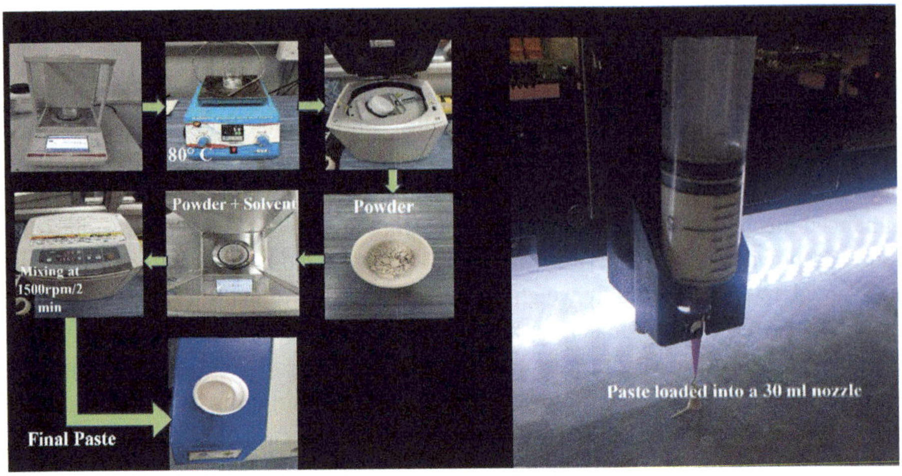

Fig. 1: Illustration of the schematic diagram of paste preparation.

Table 1: Paste composition for the DIW process.

Materials	Composition
Si_3N_4	9.0 g
Al_2O_3	0.6 g
Y_2O_3	0.4 g
HPMC	0.15 g
H_2O	20 mL
CMC	0.2 g

agent to reduce cracking during drying and stop the green body from sticking to the substrate. The platform offers a feedback speed resolution of 1.2 μm and a linear positioning accuracy of ±4 μm. Printing was done using a 0.8-mm nozzle at speeds ranging from 7 to 10 mm/s. The first layer was deposited with a thickness of 0.65 mm, followed by a second layer at 0.7 mm, with a 1 mm gap between the lines. To prevent the bottom filament from bending or drying too quickly, the printer's substrate temperature was kept steady at 10 °C. The ability of the printer to create intricate designs appropriate for mechanical testing was demonstrated by printing a variety of TPMS geometries such as gyroid, Schwarz primitive, and grid patterns with bigger pore structures. The drying rate of the extruded material increases with ambient temperature, especially affecting the bottom layers of the printed structure. Throughout the process, the substrate temperature was meticulously regulated to limit the drying time of the wet green body to prevent premature drying of these lower layers and guarantee flawless printing of all items. The optimal printing parameters for different nozzle types were

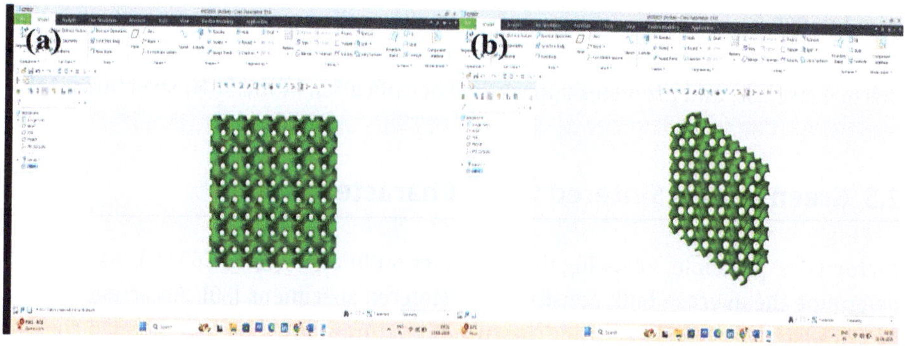

Fig. 2: 3D computer-aided design (CAD) model of a triply periodic minimal surface (TPMS) gyroid structure designed in Creo Parametric 11.0.1.0 (2024 version): (a) front view and (b) isometric view.

summarized in Table 2 based on the screw displacement, printing speed, and the internal cross-sectional area of the syringe.

2.4 Debinding and Sintering

The samples were left to air-dry naturally for at least 12 h, and no visible fractures were observed during this time. Before sintering the final components, the critical step of removing the binder (debinding) from the as-printed (green) pieces must be completed. This procedure primarily involves two steps: solvent debinding and heat debinding. First, the green body is soaked in a solvent for a specified amount of time to dissolve part of the binder and other additives. Next, thermal debinding is carried out, during which the residual binder is eliminated by vaporisation and degradation at temperatures between 200 °C and 600 °C. After the debinding procedure is finished, the components are prepared for the sintering stage. Sintering is then performed in a nitrogen atmosphere. The sintering process begins with heating to 1,200 °C at a rate of 10 °C per minute, followed by a slower increase to 1,700 °C at 5 °C per minute. The temperature is then held steady at 1,700 °C for 2 h. After sintering, it is gradually re-

Table 2: The optimum printing parameters employed in this study.

Nozzle size (mm)	0.6 mm
Layer width (mm)	0.32 mm
Layer height (mm)	0.32 mm
Print speed (mm/s)	7 mm/s
Volumetric flow rate (mm/s)	1 mm/s
Infill density	40%

duced to 800 °C at 5 °C per minute before finally cooling to room temperature inside the furnace. This regulated heat cycle ensures successful binder burnout, minimizes thermal stresses, and promotes appropriate densification of the Si_3N_4 structures [14].

2.5 Green Versus Sintered Si_3N_4: Characterization

Archimedes' principle, following the guidelines outlined in ASTM C693-93, was used to determine the average bulk density of the sintered specimens [30]. An actual density of 3.26 g/cm^3 was calculated using the rule of mixtures, based on the expected composition with sintering additives. Linear shrinkage during the HIP (sintering) process was measured by comparing the dimensions of the dried green bodies with those of the final sintered components. The microstructure of the debonded and fully sintered specimens was examined by SEM using a Nova Nano SEM 450 from FEI Company of USA (S.E.A.) PTE, Ltd. The rheological properties of the slurry were analysed by using a stress-controlled rheometer (MCR 702e MultiDrive, Anton Paar India), which was equipped with a 4 cm plate and a 1 mm gap. Oscillatory testing was carried out at a constant frequency of 1 Hz, and stress levels ranging from 10 to 5,000 Pa were applied to assess the flow behaviour under varying shear conditions.

For detailed particle size analysis, high-resolution imaging was performed using the EVO SEM (MA15/18, Carl Zeiss Microscopy Ltd.), allowing precise evaluation of the particles' shape and size, as well as their surface characteristics, key elements for understanding material performance. Mechanical testing involved a four-point bending test on as-printed TPMS gyroid structures measuring 20 × 20 × 16 mm. The tests were conducted following ASTM C1674-11 using an Instron 5881 machine (Norwood, MA) equipped with a fully articulated fixture (outer span: 20 mm; inner span: 10 mm). The samples were tested at a crosshead speed of 0.2 mm/min without any surface polishing, providing clear insights into their mechanical strength and integrity [1].

3 Results and Discussion

3.1 Rheology and Structural Characteristics of Green and Sintered 3D-Printed Si_3N_4

The morphology, particle size distribution, and zeta potential of the Si_3N_4 powder without any sintering additives, measured across different potential of hydrogen (pH) levels, are shown in Fig. 3 (a, b) and 4, respectively. The specifications provided by the manufacturer are met by the average particle size tested (D_{50} = 1.7 μm). The isoelectric point (IEP = 6.2) and the measured zeta potential curve are found to be in agreement with previously reported results for Si_3N_4 powders from the same manufacturer. Although this work did not test the IEPS of the sintering additives directly, literature studies show that high-purity Al_2O_3 usually has an IEP in the range of 8–9 and Y_2O_3 in the range of 8.5–9.5. The near-stoichiometric purity of these oxides means that their IEPs are expected to remain within these predetermined limits. When 8.0 wt.% of high-molecular-weight PVA was added to HPLC water (based on the mass of the Si_3N_4 powder), the pH of the solution was increased to 11.5. The addition of 42 vol.% of Si_3N_4 powder, along with Y_2O_3 and Al_2O_3 sintering additives, caused a slight decrease in pH to 11.4. This small change suggests that the surface adsorption of PVA played a bigger role in keeping the Si_3N_4 particles well-dispersed rather than any interaction between the powder and the surrounding solution.

Fig. 5 (a) shows how the viscosity and rheological properties of solutions containing 42 vol.% Si_3N_4 are affected by the percentage of high-molecular-weight PVA. As the PVA concentration increased from 0 to 4 wt.%, the viscosity dropped noticeably, and the flow behaviour shifted from shear-thinning to almost Newtonian. Very few additional changes were observed when the PVA concentration was raised to 8%, suggesting that the dispersant's efficacy had reached a saturation point. The impact of mixing PVA with varying molecular weights on the suspensions' rheological characteristics is seen in Fig. 5 (b). Compared to the solution without PVA, the viscosity and flow properties were not significantly altered by the addition of 4 wt.% low-M_w PVA, indicating inadequate particle dispersion. Interestingly, combining 4.0 wt.% L-PVA with 4.0 wt.% H-PVA resulted in a near-Newtonian flow profile and a significant reduction in viscosity, similar to what was observed with 4.0 wt.% H-PVA alone.

In Fig. 6, gyroid structures exceptionally engineered for mechanical testing under compression and flexure are displayed in typical green bodies made via robocasting. Both porous gyroid and solid infill architectures were produced for every specimen. The structural integrity and exact geometry of every green body were consistent. Indicating outstanding shape integrity and dimensional stability throughout the robocasting process, there was no discernible deformation, filament distortion, or collapse.

3.2 Debinding of Robocast Si_3N_4 Green Part

The thermogravimetric analysis (TGA) curves for the binder system used to make robocasted filaments are displayed in Fig. 7, with results shown both with and without the addition of Si_3N_4 powder. Before analysis, the samples were pre-dried. There were three different phases of heat deterioration in the binder-only system. Due to the evaporation of volatiles and low-molecular-weight components, there was an initial, steady weight loss (~22 wt.%) between 110 °C and 310 °C. Between 310 °C and 410 °C, the core polymer network was dramatically broken, with an additional 45–50 wt.% loss. The breakdown rate decreased at 410 °C, and even with a 4 °C/min heating rate, no binder was removed entirely, leaving residue at 600 °C. Confirming the binder's thermal behaviour inside the composite system, the TGA curve for the filament con-

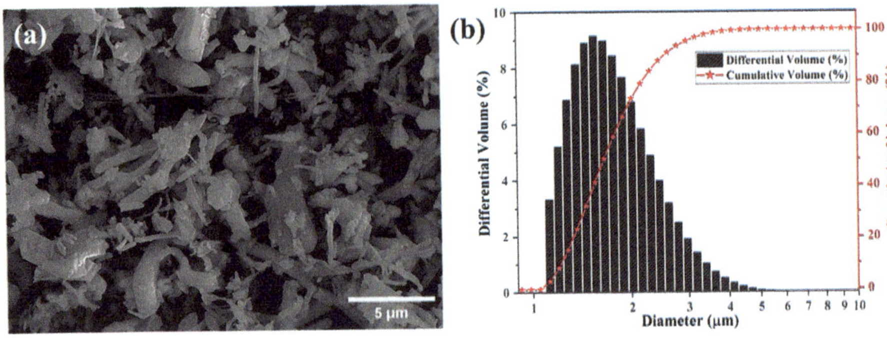

Fig. 3: (a) SEM micrograph showing the morphological characteristics and distribution of Si_3N_4 particles and (b) corresponding particle size distribution curve.

taining Si_3N_4 had a comparable deterioration trend. While the transitions were less distinct than in the binder-only system, three degradation areas were shown in the TGA curve of the filament containing Si_3N_4 powder and the binder system. The overall weight loss happened gradually and steadily throughout the entire temperature range, from room temperature up to 600 °C. The most noticeable weight loss was recorded between 310 °C and 410 °C; however, it was not as evident as it would have been for the pure binder. Interestingly, a larger percentage of weight loss was observed below 310 °C, suggesting that degradation in the first phase may have been enhanced by the presence of Si_3N_4 particles and sintering additives (Y_2O_3 and Al_2O_3) through catalysis of early-stage binder dissolution.

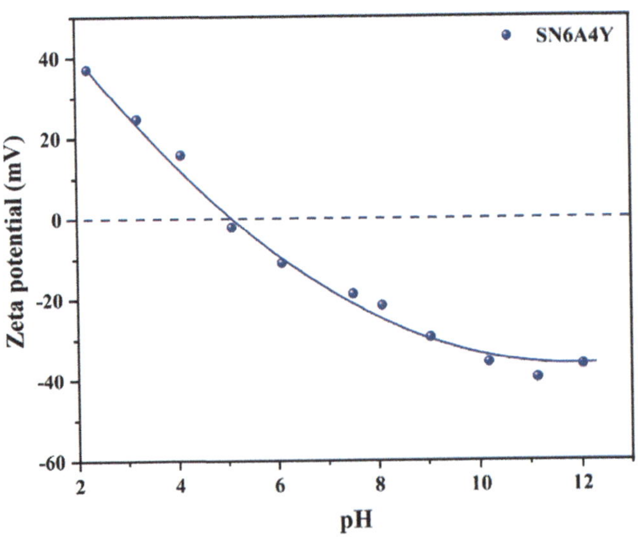

Fig. 4: Zeta potential variation with pH for as-received Si_3N_4 powder.

Following binder burnout, Fig. 8 displays SEM micrographs of a Si_3N_4 filament's surface and broken cross section taken at various magnifications. As expected, the filament surface was smooth, but the fractured cross section was rougher Fig. 8 (a). With the most significant holes measuring less than 1–2 µm in diameter, higher-magnification imaging showed equally dispersed pores Fig. 8 (b), indicating effective binder removal and the lack of notable structural flaws.

3.3 Microstructure Analysis of Sintered Si_3N_4 Specimens

Optical micrographs of the Si_3N_4 components after post-sintering and HIP are shown in Fig. 9. The final components demonstrated exceptional dimensional stability compared to the green body shown in Fig. 6. No discernible distortion in the overall geometry or filament deformation in the porous grid-like structures was observed. This shows that the printed structure kept its shape and strength throughout the sintering and HIP process. In the porous samples, an apparent density of the solid phase of 3.20 ± 0.02 g/cm^3 was measured. This corresponds to almost 98% of the theoretical density (~3.26 g/cm^3) reported for fully densified Si_3N_4 treated with the same sintering additives, demonstrating efficient densification. The solid architecture samples had a bulk density of 3.21 ± 0.02 g/cm^3, which is very close to the theoretical density of fully sintered Si_3N_4. Interestingly, the average densities measured for both the porous and solid designs did not vary much between different shapes such as cubes or beams. This suggests that the material consolidated consistently, no matter the part's geome-

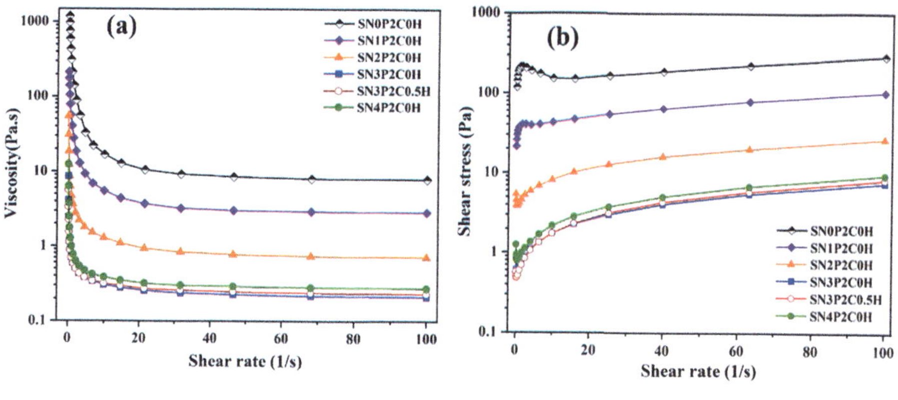

Fig. 5: (a) Viscosity-shear rate relationship for aqueous 42 vol.% Si_3N_4 suspensions with varying high-molecular-weight PVA concentrations and (b) shear stress plotted against shear rate, with the concentration of PVA shown as a percentage of the total weight of the Si_3N_4 powder.

try. A linear shrinkage of 25.8% was observed post-sintering and HIP, reflecting significant densification without compromising structural integrity. The porous grid-like Si_3N_4 structures, specifically the gyroid sample, showed an open microporosity of 50.5%, with an average pore size of around 500 µm across the deposition plane.

SEM pictures of the gyroid-structured Si_3N_4 sections with a porous architecture in the *z*-direction are displayed in Fig. 9 (a). Increased magnification revealed a small number of isolated holes in the solid portions of the microstructure, each less than 3 µm, indicating little residual porosity and efficient densification. Fig. 9 (b) demonstrated a completely dense microstructure devoid of detectable porosity, hence validating effective consolidation without contrived porosity. A fibrous microstructure made up of rod-like grains, each about 3 µm in diameter, was featured on the surface of the as-fabricated Si_3N_4 filaments with a porous grid-like structure, as revealed by SEM images. Very little glassy or solidified liquid phase was observed between the rod-shaped grains, which had a characteristic hexagonal cross section typical of β-Si_3N_4. The presence of these rod-like grains was also revealed inside the filaments by the fractured cross section, particularly in areas where intergranular fractures were more prominent. Although the grains inside the filaments seemed connected by the solidified liquid phase, their sizes were similar to those found on the surface. The as-fabricated Si_3N_4 components had an average surface roughness (R_a) of about 0.65 µm, with a root-mean-square roughness (R_q) of around 0.85 µm.

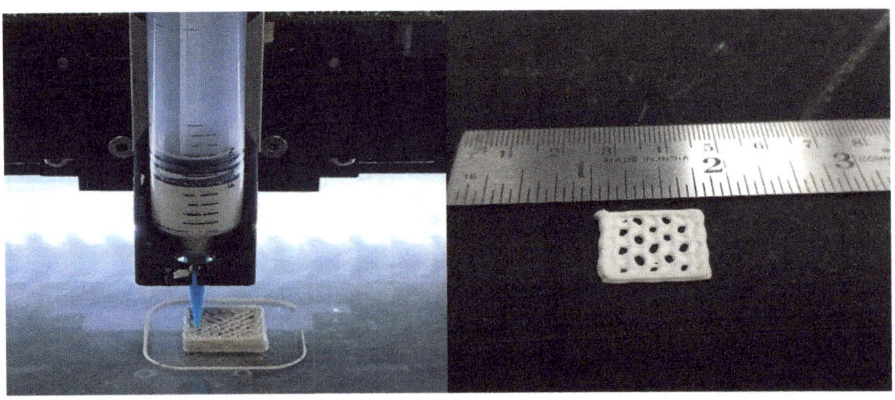

Fig. 6: Si_3N_4 green portions captured by optical means that have a solid or porous grid-like structure and gyroid-like shape.

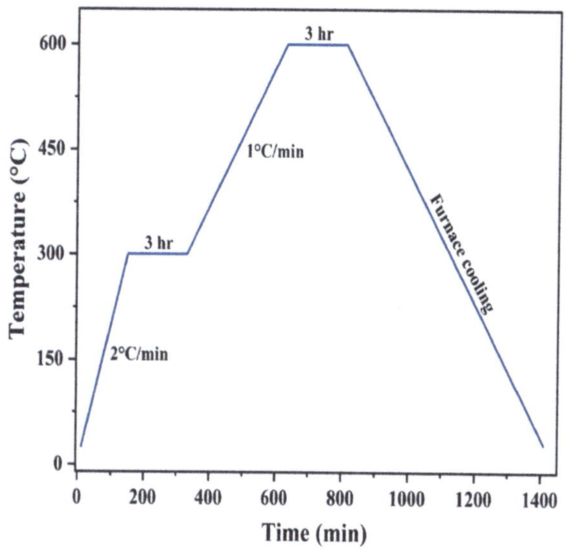

Fig. 7: The binder and paste materials were evaluated for their thermal decomposition behaviour by TGA over time.

Fig. 8: (a) SEM image shows evenly packed particles with no significant surface voids, indicating uniform packing after debinding and (b) displays uniformly spaced micro-pores (<2 μm), suggesting compact structure and efficient binder removal.

3.4 Mechanical Property Analysis of As-Fabricated Si_3N_4 Structure

At the macroscopic level, the mechanical properties of grid-structured specimens are strongly influenced by factors such as filament diameter, as well as the size and shape of the macroscopic pores. First, the effect of macropore size on mechanical characteristics and relative density was investigated. With a printing spacing of 0.4 mm, the printed sample achieved an impressive relative density of 98%, reflecting high material integrity and minimal porosity. This was based on the SEM pictures in Fig. 7, the printed materials' pore size increases as the printing gaps widen, which affects the flexural strength. Since it was more than the mechanical testing machine's load cell's maximum capacity (10 kN), the gyroid-structured part's compressive strength could not be determined. This indicates that the porous grid-like samples, featuring a macroporosity of 50.5%, demonstrated a compressive strength exceeding 200 MPa, highlighting their strong structural integrity despite their porous design. Compressive strength is projected to be substantially higher in fully dense components with a solid design. Four-point bending tests revealed that the solid sections of the as-fabricated gyroid achieved a flexural strength of 552 ± 68 MPa, while the grid-like porous parts with 50.5% microporosity recorded a flexural strength of 85 ± 12 MPa. SEM examinations of the cracked surfaces after testing revealed intergranular and trans granular fracture processes. Fewer intergranular fractures were observed in the upper region under compression, whereas more were displayed in the bottom region under tension. Despite these variations, the same breakage patterns were shown by both solid and porous structures.

3.5 Progress in Si3N4 for DIW

The successful and continuous robocasting of green components was enabled by the development of an aqueous-based paste formulated with fine commercial Si_3N_4 powder, standard sintering additives (Y_2O_3 and Al_2O_3), and organic processing agents. This innovative paste allowed for the creation of components with a wide range of geometries and architectural designs, ensuring both versatility and precision in fabrication. Through this manufacturing process, precise layering of the material was achieved, resulting in well-formed green bodies that are ideal for further densification. Further densification was achieved by combining sintering with sinter/HIP, which improved material consolidation while ensuring that the desired shape was maintained without noticeable deformation. The final Si_3N_4 components exhibited a nearly fully dense microstructure with rod-like grains, indicating good material integrity and homogeneity.

Mechanical evaluation through four-point bending tests demonstrated that the fabricated dense samples achieved an average flexural strength of 552 MPa, showcasing their impressive robustness and reliability after processing. The strength values obtained are auspicious for load-bearing applications where mechanical reliability is essential. The described fabrication method provides a reliable and scalable route for producing Si_3N_4 implants with intricate designs and anatomically relevant features. Structural integrity during densification is effectively maintained, and high flexural strength is achieved, making this method a strong candidate for orthopaedic and dental applications, where complex architectures and biocompatibility are vital. Manufacturing precision is enhanced, and this approach opens new pathways for creating customised medical implants tailored to patient-specific anatomical requirements.

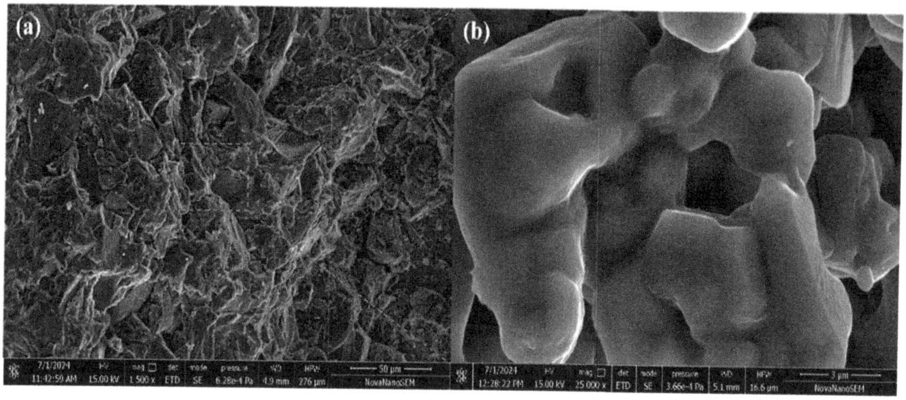

Fig. 9: Polished cross sections of the sintered Si_3N_4 parts are displayed in the SEM images, with two distinct structures being highlighted: (a) a porous grid-like architecture and (b) a solid architecture, both shown in greater detail at higher magnification.

4 Conclusions

To facilitate the continuous robocasting of green components with a variety of shapes and topologies, an aqueous paste including Si_3N_4 powder, sintering additives (Y_2O_3 and Al_2O_3), and organic processing agents was created. Using this technique, green components ideal for densification later on were precisely and consistently fabricated. After completing a two-step sintering process followed by sinter/HIP, two distinct types of components were obtained: solid parts with nearly full-density Si_3N_4, boasting a density of 3.21 ± 0.02 g/cm^3, and macroporous sections featuring a grid-like structure. The fibrous grain microstructure, characteristic of in situ-toughened Si_3N_4, was responsible for the improved mechanical qualities exhibited by the solid portions.

Mechanical testing of the as-fabricated solid samples, designed in a beam-like shape, was performed using four-point bending tests. The results revealed an impressive average flexural strength of 552 ± 68 MPa, indicating excellent material robustness. This strong performance suggests that the samples are well-suited for load-bearing applications. The macroporous grid-like sections, characterized by a 500-µm pore diameter and 50.5% porosity, displayed a flexural strength of 85 ± 12 MPa, which is notably lower compared to their solid counterparts. Despite this difference, the results highlight that robocasting combined with sinter/HIP serves as an effective approach for fabricating Si_3N_4 components with complex internal architectures. This manufacturing technique shows great promise for producing customized Si_3N_4-based implants with tailored internal designs and anatomically accurate shapes. The findings suggest that this method could be a viable pathway for creating personalized implants for dental and orthopaedic applications, where detailed geometries and reliable mechanical performance are crucial.

References

[1] Zhao S, Xiao W, Rahaman MN, O'Brien D, Seitz-Sampson JW, Sonny Bal B. Robocasting of Silicon Nitride with Controllable Shape and Architecture for Biomedical Applications. Int J Appl Ceram Technol. 2017 Mar 1;14(2):117–27.

[2] Klemm H. Silicon Nitride for High-temperature Applications. J Am Ceram Soc. 2010 Jun;93 (6):1501–22.

[3] Dai YL, Yao DX, Xia YF, Zhu M, Zhao J, Wang F, et al. Preparation of Porous Si3N4 Ceramics with Varied Macropore Structures by Direct Ink Writing. Ceram Int. 2024 Dec 1;50(23):49033–49040.

[4] Pezzotti G, Marin E, Adachi T, Rondinella A, Boschetto F, Zhu W, et al. Bioactive Silicon Nitride: A New Therapeutic Material for Osteoarthropathy. Sci Rep. 2017 Mar 22;7:44848.

[5] Becher PF, Sun EY, Plucknett KP, Alexander KB, Hsueh CH, Lin HT, et al. Microstructural Design of Silicon Nitride with Improved Fracture Toughness: I, Effects of Grain Shape and Size. J Am Ceram Soc. 1998 Mar 17; 81(11):2821–30.

[6] Sainz MA, Serena S, Belmonte M, Miranzo P, Osendi MI. Protein Adsorption and in Vitro Behavior of Additively Manufactured 3D-silicon Nitride Scaffolds Intended for Bone Tissue Engineering. Mater Sci Eng C. 2020 Oct 1;115:110734.

[7] Rahaman MN, Yao A, Bal BS, Garino JP, Ries MD. Ceramics for Prosthetic Hip and Knee Joint Replacement. J Am Ceram Soc. 2007 Jul;90(7):1965–88.

[8] Bal BS, Garino J, Ries M, Rahaman MN. Ceramic Materials in Total Joint Arthroplasty. JSESArthro. 2006;17(3–4):94–101.

[9] Millaâ AJ, Nieto MI, Moreno R. Aqueous Injection Moulding of Silicon Nitride. Journal of the European Ceramic Society. 2000 Dec; 20(14-15):2661–2666.

[10] Albano MP, Garrido LB. Processing of Concentrated Aqueous Silicon Nitride Slips by Slip Casting. J. Am. Ceram. Soc. 1998 Apr; 81(4):837–44.

[11] Li S, Sassa K, Asai S. Preferred Orientation of Si3N4 Ceramics by Slip Casting in a High Magnetic Field. Ceram Int. 2006;32(6):701–05.

[12] Wan T, Yao D, Hu H, Xia Y, Zuo K, Zeng Y. Fabrication of Porous Si3N4 Ceramics through a Novel Gelcasting Method. Mater Lett. 2014 Oct; 15;133:190–92.

[13] Wu JM, Zhang XY, Yang JL. Novel Porous Si3N4 Ceramics Prepared by Aqueous Gel casting Using Si3N4 Poly-hollow Microspheres as Pore-forming Agent. J Eur Ceram Soc. 2014 May;34(5):1089–96.

[14] Shahzad A, Lazoglu I. Direct Ink Writing (DIW) of Structural and Functional Ceramics: Recent Achievements and Future Challenges. Composites Part B: Engineering. Elsevier Ltd 2021;225:109249.

[15] Kotobuki M. Properties of Al2O3 Pastes Using Inorganic Na2SiO3 Binder and Organic Binder for Direct Ink Writing. Phys Status Solidi B Basic Res. 2022 Sep 1;259(9):2100520.

[16] Zhu K, Yang D, Yu Z, Ma Y, Zhang S, Liu R, et al. Additive Manufacturing of SiO2–Al2O3 Refractory Products via Direct Ink Writing. Ceram Int. 2020 Dec 1;46(17):27254–61.

[17] Costakis WJ, Wyckoff C, Schlup A, Wallace M, Craigs T, Malek E, et al. Material Extrusion of Highly-loaded Silicon Nitride Aqueous Inks for Solid Infilled Structures. Addit Manuf. 2023 Feb 25;64:103425.
[18] Diener S, Franchin G, Achilles N, Kuhnt T, Rösler F, Katsikis N, et al. X-ray Microtomography Investigations on the Residual Pore Structure in Silicon Nitride Bars Manufactured by Direct Ink Writing Using Different Printing Patterns. Open Ceramics. 2021 Mar 1;5:100042.
[19] Rueschhoff LM, Youngblood JP, Trice RW. Stabilizing Highly Loaded Silicon Nitride Aqueous Suspensions Using Comb Polymer Concrete Superplasticizers. J Am Ceram Soc. 2016;99(12):3857–65.
[20] Rueschhoff LM, Trice RW, Youngblood JP. Near-net Shaping of Silicon Nitride via Aqueous Room-temperature Injection Molding and Pressureless Sintering. Ceram Int. 2017 Oct 1;43(14):10791–98.
[21] Lewis JA, Smay JE, Stuecker J, Cesarano J. Direct Ink Writing of Three-dimensional Ceramic Structures. J Am Ceram Soc. 2006 Dec;89(12):3599–609.
[22] Feilden E, Blanca EGT, Giuliani F, Saiz E, Vandeperre L. Robocasting of Structural Ceramic Parts with Hydrogel Inks. J Eur Ceram Soc. 2016 Aug 1;36(10):2525–33.
[23] Miranda P, Pajares A, Saiz E, Tomsia AP, Guiberteau F. Mechanical Properties of Calcium Phosphate Scaffolds Fabricated by Robocasting. J Biomed Mater Res A. 2008 Apr;85(1):218–27.
[24] Miranda P, Saiz E, Gryn K, Tomsia AP. Sintering and Robocasting of β-tricalcium Phosphate Scaffolds for Orthopaedic Applications. Acta Biomater. 2006;2(4):457–66.
[25] Maazouz Y, Montufar EB, Guillem-Marti J, Fleps I, Öhman C, Persson C, et al. Robocasting of Biomimetic Hydroxyapatite Scaffolds Using Self-setting Inks. J Mater Chem B. 2014 Sep 7;2 (33):5378–86.
[26] Fu Q, Saiz E, Tomsia AP. Direct Ink Writing of Highly Porous and Strong Glass Scaffolds for Load-bearing Bone Defects Repair and Regeneration. Acta Biomater. 2011;7(10):3547–54.
[27] Liu X, Rahaman MN, Hilmas GE, Bal BS. Mechanical Properties of Bioactive Glass (13–93) Scaffolds Fabricated by Robotic Deposition for Structural Bone Repair. Acta Biomater. 2013;9(6):7025–34.
[28] Xiao W, Zaeem MA, Bal BS, Rahaman MN. Creation of Bioactive Glass (13–93) Scaffolds for Structural Bone Repair Using a Combined Finite Element Modeling and Rapid Prototyping Approach. Mater Sci Eng C. 2016 Nov 1;68:651–62.
[29] He GP, Hirschfeld DA, Cesarano Iii J, Stuecker JN ROBOCASTING AND MECHANICAL TESTING OF AQUEOUS SILICON NITRIDE SLURRIES ~ % C~.
[30] Test Method for Density of Glass by Buoyancy [Internet]. West Conshohocken, PA: ASTM International; 2008. Available from: http://www.astm.org/cgi-bin/resolver.cgi?C693-93R08
[31] Kumar A, Rajak DK, Kumar P, Kumar A, editors. Advances in Sustainable Biomaterials: Bioprocessing 4.0, Characterizations, and Applications. CRC Press; 2024 Oct 28.
[32] Anand S, Satyarthi MK, Bharti PS, Kumar P, Kumar A. Digital Twin Integration for Enhanced Control in FDM 3D Printing. In: Industry 4.0 Driven Manufacturing Technologies. Cham: Springer Nature Switzerland;2024 Sep 14. 373–88.
[33] Yadav P, Kumar P, Saxena M, Dwivedi Y, Dubey A. 15 Application of Three-dimensional Printing in Medical, Agriculture, Engineering, and Other Sectors. 3D Printing Technologies: Digital Manufacturing, Artificial Intelligence, Industry 4.0. 2024 Jan;29:311.
[34] Sellamuthu P, Kalita K, Kumar A, Chohan JS, Rathee S, Sharma YK. A hybrid Merec-Mabac approach for 3D printer nozzle material selection. In: AIP Conference Proceedings. AIP Publishing; 2024 Dec 20. 3217, No. 1.
[35] Kumar A, Kumar P, Sebaey TA, Gambhir V, editors. Tools, Techniques, and Advancements in Engineering Materials Science. IGI Global; 2024 Dec 24.

List of Contributors

Chapter 1
Amandeep Singh
Research Scientist
India Institute of Technology (IIT)
Ropar
Punjab 140001
India

Sarthik Samanta
Earth Science and Engineering
Missouri University of Science and Technology
Missouri
Rolla, MO 65409
USA

Santosh K. Tiwari
State Key Laboratory of Featured Metal Materials and Life-cycle Safety for Composite Structures &
MOE Key Laboratory of New Processing Technology for Nonferrous Metals and Materials
School of Resources, Environment and Materials
Guangxi University
Nanning 530004
China
Centre for New Materials and Surface Engineering
NMAM Institute of Technology
Nitte
Karnataka 574110
India

Chapter 2
Parveen Kumar
Skill Department of Automotive Studies
Shri Vishwakarma Skill University
Dudhola
Plawal-121102
Haryana
India

Devesh Kumar Pandey
Department of Civil Engineering
Rawal Institute of Engineering & Technology
Faridabad-121004
Haryana
India

Mansish Tomar
Department of Civil Engineering
Rawal Institute of Engineering & Technology
Faridabad-121004
Haryana
India

Gaurav Sharma
Department of Mathematics
Rawal Institute of Engineering & Technology
Faridabad-121004
Haryana
India

Rajeev Sindhu
Department of BCA & B.Sc. (Computer Science)
GNIOT Institute of Professional Studies
Knowledge Park-2
Greater Noida-201310
Uttar Pradesh
India

Chapter 3
Erukala Kalyan Kumar
Department of Mechanical Engineering
National Institute of Technology
Rourkela, 769008
Odisha
India
Department of Mechanical Engineering
Presidency University
Itgalpur
Rajanakunte
Bengaluru, 560064
Karnataka

Chapter 4

SHUBHRA Shitole
Mechanical and Aerospace Engineering
Department
College of Engineering
United Arab Emirates University
Al-Ain, 15551
United Arab Emirates

S. Vincent
Department of Mechanical Engineering
BITS Pilani
Dubai Campus
Dubai 345055
United Arab Emirates

ABDEL-HAMID Ismail Mourad
Mechanical and Aerospace Engineering
Department
College of Engineering
United Arab Emirates University
Al-Ain, 15551
United Arab Emirates
National Water and Energy Center
United Arab Emirates University
Al Ain, 15551
United Arab Emirates

PRIYA Sudha
Department of Mechanical Engineering
BITS Pilani
Dubai Campus
Dubai 345055
United Arab Emirates

Chapter 5

Ruthviz, Kodali
Ecole Centrale College of Engineering
Mahindra University
Hyderabad
Telangana, 500043
India

Semanth Reddy Bommu
Ecole Centrale College of Engineering
Mahindra University
Hyderabad
Telangana, 500043
India

Abhishek Bondada
Ecole Centrale College of Engineering
Mahindra University
Hyderabad
Telangana, 500043
India

Jayaprakash Vemuri
Ecole Centrale College of Engineering
Mahindra University
Hyderabad
Telangana, 500043
India

Chapter 6

Parveen Kumar
Skill Department of Automotive Studies
Shri Vishwakarma Skill University
Dudhola
Plawal-121102
Haryana
India

Devesh Kumar Pandey
Department of Civil Engineering
Rawal Institute of Engineering & Technology
Faridabad-121004
Haryana
India

Mansish Tomar
Department of Civil Engineering
Rawal Institute of Engineering & Technology
Faridabad-121004
Haryana
India

Gaurav Sharma
Department of Mathematics
Rawal Institute of Engineering & Technology
Faridabad-121004
Haryana
India

Rajeev Sindhu
Department of BCA & B.Sc. (Computer Science)
GNIOT Institute of Professional Studies
Knowledge Park-2
Greater Noida-201310
Uttar Pradesh
India

Chapter 7
G Naga Srinivasulu
Department of Mechanical Engineering
National Institute of Technology Warangal
Hanamkonda- 506004
Telangana
India

N V Raghavaiah
Department of Mechanical Engineering
National Institute of Technology Warangal
Hanamkonda- 506004
Telangana
India

Chapter 8
Ahsan Ul Haq
Department of Mechanical Engineering
BITS Pilani
Hyderabad Campus
Hyderabad
Telangana, 500078
India

Suresh Kumar Reddy Narala
Department of Mechanical Engineering
BITS Pilani
Hyderabad Campus
Hyderabad
Telangana, 500078
India

Chapter 9
Katika Harikrishna
Department of Mechanical Engineering
National Institute of Technology-Warangal-
506004
India
Department of Mechanical Engineering
Shri Vishnu Engineering College for Women
Bhimavaram 534202
India

Maharajan Joseph Davidson
Department of Mechanical Engineering
National Institute of Technology-Warangal
506004
India

Gottim Dhanush Reddy
Department of Mechanical Engineering
National Institute of Technology-Warangal
506004
India

Poorna Jagadeesh
Department of Mechanical Engineering
National Institute of Technology-Warangal
506004
India

Vasista Kadali
Department of Mechanical Engineering
National Institute of Technology-Warangal
506004
India

Shaik Naveed Ahammed
Department of Mechanical Engineering
National Institute of Technology-Warangal
506004
India

Chapter 10
Gaddam Ashok
Department of Mechanical Engineering
SR University
Warangal 506371
India

Pankaj kumar
Department of Mechanical Engineering
SR University
Warangal 506371
India

Ajay Kumar
Department of Mechanical Engineering
School of Core Engineering, Faculty of Science, Technology & Architecture, Manipal University
Jaipur
Rajasthan
India

Chapter 11

Ruchira Srivastava
Department of Mechanical Engineering
Amity University
Greater Noida-201310
Uttar Pradesh
India

Ayushi Thakur
Department of Mechanical Engineering
Amity University
Greater Noida-201310
Uttar Pradesh
India

Parveen Kumar
Skill Department of Automotive Studies
Shri Vishwakarma Skill University
Dudhola
Plawal-121102
Haryana
India

Ajay Kumar
Department of Mechanical Engineering
School of Core Engineering, Faculty of Science, Technology & Architecture, Manipal University
Jaipur
Rajasthan
India

Chapter 12

Narinder Kumar
Department of Mechanical Engineering
Chandigarh Engineering College-CGC
Landran
Mohali
India

Hardeep Singh
Department of Mechanical Engineering
Chandigarh Engineering College-CGC
Landran
Mohali
India

Manjit Singh
Department of Mechanical Engineering
Chandigarh Engineering College-CGC
Landran
Mohali
India

Chapter 13

Amit Kumar
Centre for Advanced Studies
Dr. A.P.J. Abdul Kalam Technical University
Lucknow, 226031
Uttar Pradesh
India

Siddharth Yadav
Centre for Advanced Studies
Dr. A.P.J. Abdul Kalam Technical University
Lucknow, 226031
Uttar Pradesh, India

Anuj Kumar Sharma
Centre for Advanced Studies
Dr. A.P.J. Abdul Kalam Technical University
Lucknow, 226031
Uttar Pradesh
India

Santosh Kumar
Department of Mechanical Engineering
Indian Institute of Technology (BHU)
Varanasi, 221005
Uttar Pradesh
India

Govind Kumar Verma
Department of Mechanical Engineering
Indian Institute of Technology (BHU)
Varanasi, 221005
Uttar Pradesh
India

Index

https://doi.org/10.1515/9783111515717-015

Printed and bound by CPI Group (UK) Ltd, Croydon, CR0 4YY
13/09/2026
14961094-0001